The Water Mills of the Water of Leith

Graham Priestley

the
WATER
of LEITH
Conservation Trust

Published by
The Water of Leith Conservation Trust,
24 Lanark Road,
Edinburgh EH14 1TQ

Illustrations by the author
except where credited otherwise.

Copyright © Graham Priestley, 2001
All rights reserved.

ISBN 0-9540006-0-9

Printed by
Bridgend Printers (Edinburgh) Ltd.,
40 Constitution Street, Leith,
Edinburgh EH6 6RS

CHAPTER 1. Introduction

The sight and sound of a water wheel turning, the water flowing onto it and splashing off the other side, the creaks and groans of the axle, still fascinate people today. They regard the water wheel as an intriguing oddity, a picturesque and romantic survival of a simpler world. Now, merely pressing a switch or a button will give instant light and heat, move us from place to place, or initiate rapid programmes of work by all manner of increasingly complex machines: civil engineers can literally move or even build mountains or tunnel through them to the other side.

When such things are taken for granted, it is difficult for us to imagine what life was like 1000 years ago, long before the harnessing of steam and the internal combustion engine, but we know that it must have been harder and far less comfortable. Power to do even simple things, to grind corn for bread, to spin thread and weave cloth for clothing or chop wood for fuel, was usually then a matter of bending backs and straining human muscle, with the heavier jobs reserved for horses. For tasks too big or too sustained even for horses, the only answer was a windmill or a water mill, which would do the work of six, or a dozen, or even more horses with no food or fuss - power, as it still seems to the bystander, for nothing and going on for ever. Scotland had some windmills, including one or two in Edinburgh and Leith, but the abundance of flowing water made this the usual choice and for at least 800 hundred years, well into the 19th century, local water mills provided a focus for local life. Their chosen sites below a bend in the river or at some natural waterfall often determined where settlements would develop and mills were a major influence on the growth and prosperity of many villages and towns, including the city of Edinburgh.

For the purpose of this book, a mill is simply a building or set of buildings with a water-driven engine, usually a wheel, powering machinery for some particular task. The first water mills ground corn; later mills fulled cloth - the process by which cloth was beaten and softened to finish it and make it pleasant to wear. Yet the same wheels on the same sites could be adapted for many other agricultural and industrial tasks and many mills have a long history of different uses as one product became more profitable than another or a new owner had new enthusiasms. Mill stones, of course, would grind a whole range of grains, seeds and spices. The basic machinery for paper-making, a stamper for beating the rags into a pulp, closely resembled the stocks or presses used in fulling cloth, and the trip hammers used to flatten metal in forges: in each case a series of hammer-like heads were raised by projecting cams on a rotating shaft driven by the water wheel. Fulling mills (called waulk mills in Scotland) often became paper mills in the 17th and 18th centuries.

Edinburgh's river, as we think of the Water of Leith today, is remarkable for the long history of its mills, for their number and varied uses, and for their important contribution to the development of Edinburgh and its hinterland. In England, King William I's Domesday Book of 1086 recorded over 5,000 mills, and many of these same sites survive today. Scotland had no equivalent survey but royal charters to monasteries provide a piecemeal record of the existence of some early mills, including several on the Water of Leith. In 1128 King David I granted the mills at Bonnington, Canonmills, Dean and Saughton to the Abbot of Holyrood in Edinburgh. In the Statistical Account of Scotland of 1791 there were said to be 76 mills on the river - a remarkable density for a waterway only 23 miles in its entire length and one possibly unique in Scotland, though the River Leven in Fife boasted 46 mill sites along about 11.2 miles in 1830 according to data in John Shaw's *Water Power in Scotland*.

Some Leith sites had been abandoned by the 19th century and a survey of water mills on the main river in 1850 - neglecting the important tributary Bavelaw Burn at Balerno - listed only 37 mills, though some of the sites had more than one use and more than one operator at the time and could reasonably be classified as two or even three mills. These 37 mills employed a total of 66 wheels for grinding corn and flour, paper-making, grinding snuff and spices, washing and waulking (fulling) cloth, grinding bark and pumping water for tanning, cutting and polishing ornamental stone and sawing timber. Paper-making developed into an important industry following the establishment of Scotland's first paper mill on the river at Dalry (Murrayfield) in 1595. The needs of the paper mills for coal and raw materials brought the railway up the valley to Balerno in 1874, by which time the mills at Colinton, Juniper Green, Currie and Balerno together employed hundreds of people.

Today, sadly, not a single water-powered mill exists on the river, though two mill buildings, Kinauld Tannery at Currie and Inglis Grain Mill beside the City Bypass at Juniper Green, retain something of their earlier usage and other mills survive after conversion into private dwellings. Many relics of water-powered industry can be seen along the river, however, as weirs, sluices, lades, millstones, mill houses and even a solitary wheel, and it is the purpose of this book to explain their presence and to whet the reader's appetite towards more interest in our river and in Scotland's working water mills elsewhere. Information scattered through many different publications and other varied sources is collected together for the first time into a single comprehensive account of the river's mill history. The relics decline each decade and the uncontrolled regeneration of riverside trees makes it increasingly hard to see what survives. A preservation campaign is an urgent need.

The first task is to explain how a mill works (Chapter 2) and then show which bits and pieces survive up and down the Water of Leith (Chapter 3). Chapter 4 describes some of the industries involved while Chapter 5 gives brief details of all the Water of Leith mills from the top of the river down to Leith. Appendices to the main text show the sources used in preparing this account, provide a glossary of technical terms and list some of Scotland's working water mills open to the visitor.

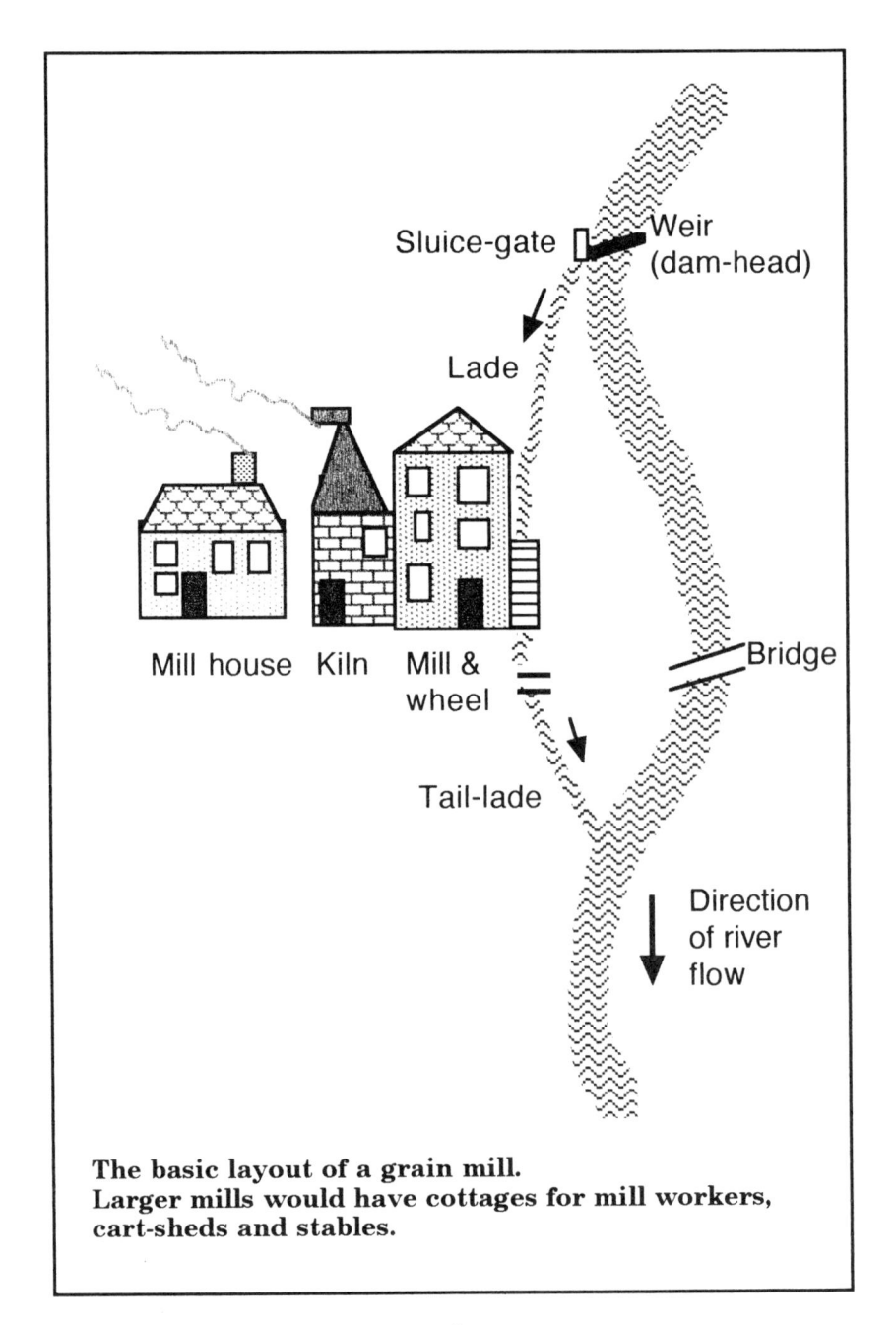

The basic layout of a grain mill.
Larger mills would have cottages for mill workers, cart-sheds and stables.

CHAPTER 2. How a mill works.

Any source of flowing water will drive a wheel of some size, though if the supply is tiny it may be necessary to store it up in a pond until there is sufficient for a few hours' work. This was seldom necessary on the Water of Leith, though summer droughts often limited working hours and in the 1850s, when many of the moorland springs at the top of the river were tapped for Edinburgh's drinking water supply, three compensation reservoirs were constructed, Harlaw, Threipmuir and Harperrig, to guarantee a workable flow down the river for the mills. The reservoirs not only maintained the river levels in summer by releasing water stored up during the winter, but also gave a degree of protection against flood damage by absorbing some of the sudden rainfall during the winter months.

Given the flow of water, the miller or millwright, the name given to a mill engineer, had to direct the water towards his mill. This was done by, in effect, building a wall across the river, usually slanting across towards the miller's own bank so that much of the flow was diverted into a channel leading to the mill while the excess spilled over into the main river bed downstream. We usually call these walls weirs, but they are also known as dams, or, on the Water of Leith, damheads. They could be massive structures 8, 10 or 12 feet high, raising the level of the water in the river upstream by the same height. Entry of water into the channel, usually called a lade (or, occasionally on the Water of Leith, the mill dam) but in other parts of Britain a cut, race, mill stream, goit, leat or flume, was controlled by a sluice-gate, usually a heavy wooden gate held at either side in grooved stonework so that it could be raised by means of a crank to release the flow of water or lowered to block it completely.

Some lades were lined with stonework but others were simply earthen ditches that required regular cleaning and dredging. The lade was engineered to slope downwards only slightly towards the mill, while the river followed a gradient, creating a difference in level, a head of water to act on the wheel. The height of the head was one factor determining the size (height) of the wheel that could be used on the site. The width of the wheel would match the width of the lade and the volume of water flowing down it - the greater the flow the wider and more powerful the wheel. At the end of the lade would be another sluice or perhaps two. One would close off the flow onto the wheel and the other, a relief or bypass sluice, would divert any water still flowing down the lade back into the river. There could be no brake on a water wheel, so if its water supply was turned on the wheel turned and the mill machinery was likely to turn with it, unless there was a mechanism for taking the machinery out of gear. Such a mechanism was provided by slipping belts and spindles or a rising jack ring to disengage the stone nuts in a grain mill.

Water wheels were once made entirely of wood but this came to be replaced by metal in the 18th century. There were several patterns of wheel. A primitive type is the Norse wheel which operated horizontally, turning on a vertical shaft. A few of these wheels survive in Orkney, Shetland, Scandinavia and France. They needed only a small water source and drove a single pair of mill stones directly, without any gearing. Such wheels may have had more widespread use but they were displaced by the next type, the undershot wheel. At its simplest this was a series of wooden paddles projecting radially from the horizontal axle like the spokes of a wheel (Fig. A opposite). Each paddle would fill the cross-section of the lade, so that all the force of the flow acted on the paddle to push the wheel and as little water as possible passed round the sides and was wasted. The wheel was turned by the force (speed and volume) of the water acting on its lower half and passing below (under) the level of the axle.

An alternative to this, requiring a greater head of water but less volume, was the overshot wheel (Fig. C). Here the water passed over the top of the wheel, filling buckets made by setting angled paddles between the two rims of the wheel and backing them with an inner plate. The wheel was

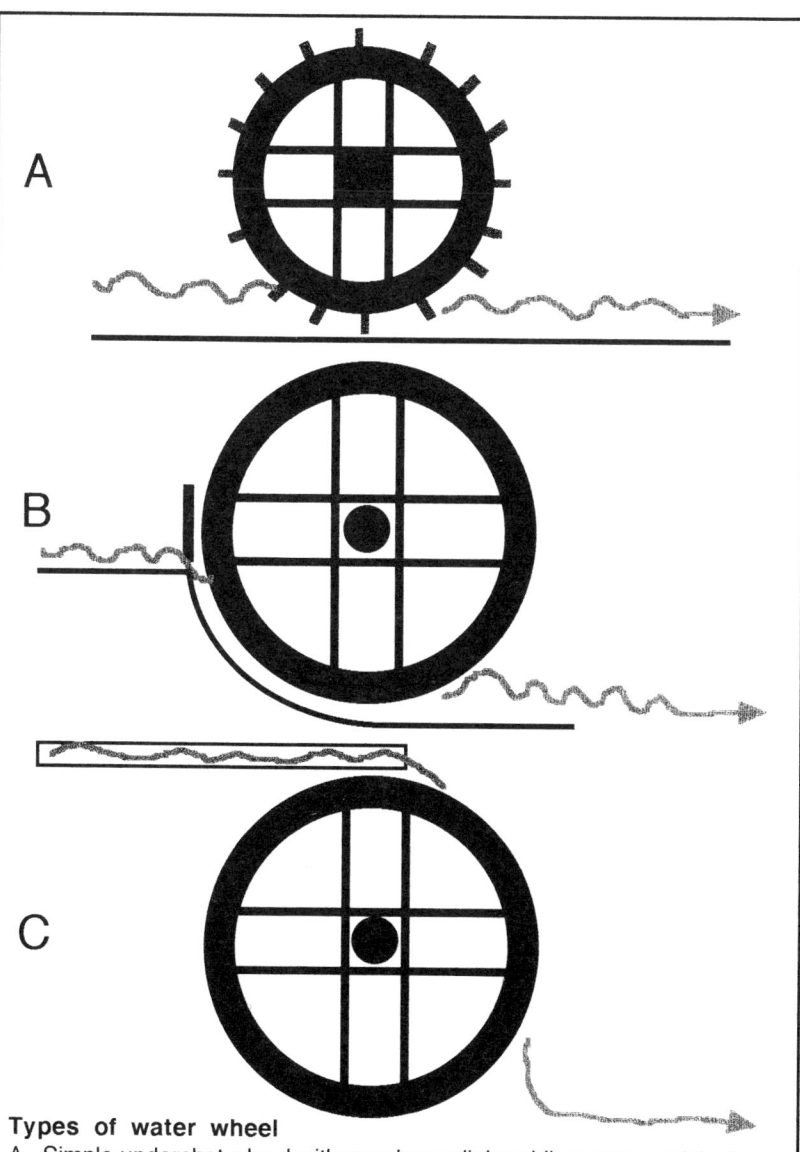

Types of water wheel
A. Simple undershot wheel with wooden radial paddles; turns anticlockwise.
B. Breast wheel with curved paddles between iron rims forming buckets; turns anticlockwise in shaped wheel pit. C. Overshot wheel with enclosed buckets, turns clockwise.

turned by gravity, by the weight of water on the forward side of the wheel. Overshot wheels could be narrow if the water supply was limited and the height of the wheel could be increased to compensate for this and produce more power. The greater head required to supply a higher wheel could be provided by lengthening the lade and so increasing the difference in level between lade and river. If the miller didn't have enough land for a long lade, or it was too flat, an overshot wheel might be impossible. On the Water of Leith, where mills were closely spaced, most wheels were undershot. Another consequence of crowding and co-existence was that neighbouring mills on the same bank of the river could share the same lade, the water being used by each undershot wheel in turn. There were several examples of this on the Water of Leith.

The Yorkshire engineer John Smeaton (1724-1792), who visited Dalry Mills in 1771 to advise on the choice of water wheels there, had earlier measured the efficiency of the different types of wheel. Undershot wheels did poorly at less than 30% and overshot wheels were much superior at 63%. Smeaton introduced several refinements to water wheels, including the use of iron for many of their components. Many undershot wheels were replaced by breast wheels (Fig. B), which were a compromise between the two types, the water entering the wheel midway, at about the level of the axle, (the shaft on which the wheel turned) and acting by gravity. These wheels had intermediate efficiency and were suitable for most sites because they needed only a small head of water. The position of a breast wheel on a lade can be recognised by its wheel pit, a sudden increase in the depth of the lade, shaped in stone to fit one quarter of the wheel closely, so that the dimensions of the wheel can still be determined even if the wheel itself is long gone. The water leaving the wheel rarely returned directly to the river. Instead it usually flowed down a tail-lade before joining the main stream.

Although there are infinite variations in the design and workings of mills and the solutions to problems devised by generations of millwrights, according to the particular site and the product, many features follow a common pattern. Wheels could be positioned inside or outside the mill. If outside, they might be protected from the weather in a wheel-house, the axle of the wheel passing into the mill itself through a hole in the wall to drive the machinery.

For a grain mill, the drive had to be transmitted from the vertical to the horizontal plane to operate the millstones (see page 23). On the inside of the wall there was a large pit wheel, so named because it required a deep pit. Its teeth would enmesh with those of a horizontal wheel (the trundle) which transferred the drive to a vertical shaft. This shaft bore the spur wheel which transmitted the drive to pinions and secondary vertical shafts driving individual pairs of millstones on the floor above. In each pair of stones the lower (bed) stone remained stationary and the rotating shaft passed through it to turn the upper (runner) stone with a carefully adjusted gap between the two stones in which the grain was crushed. The runner stone turned at 150-250 revolutions per minute depending on its diameter.

The working surfaces of the two stones were usually dressed with a special pattern of grooves or furrows, shallow at the eye (centre) of the stone and deeper at the edge. The dressings on opposed surfaces of the two stones were mirror images so that the sharp edges of the furrows cut the grain like scissors. The crushed meal or flour travelled by centrifugal force along the furrows to fall off the outer rim of the stone and be collected via a chute. In the commonest dressing the surface of the stone was divided into 10 identical segments, each containing four parallel furrows aligned tangentially to the eye of the stone. A simplified version of this pattern, with only 6 segments, has been adopted by the Water of Leith Conservation Trust as its logo.

Grain mills were usually three or more storeys high. The grain was hoisted up to the top floor where it was stored and fed down a chute into the eye of the stones on the floor below. From the stones the meal or flour was swept into another chute and fell down to a lower floor for bagging and transfer to carts at ground level. Ancillary uses of the water wheel were to power a sack hoist so that the grain could be lifted up to the top floors and to work sieves, sometimes called boulting machines, to sift the flour. After about 1800 some mills had a steam engine, at first to help out at times of drought, and then to supplement or replace the water wheel, so that tall chimneys were added to the landscape. The early steam engines were unreliable and expensive and were thought to increase the ever-present risk of fire.

The mill needed constant attention and the miller had to live nearby, so there is usually a mill house beside the mill, perhaps with cottages for the mill hands, cart sheds and stables. Some sites had two houses: a grand house for the mill owner, as at Stenhouse and Roseburn, and a simpler house for the miller, his tenant. Grain had to be dry for milling, so many mills had their own kiln where the grain was spread thinly on an upper floor, perhaps of perforated tiles supported in an iron framework, to be dried by heat from a fire on the floor below. Kilns had a tapered roof topped by a cowl to direct the smoke away and were an attractive feature of the mill scene. There would be a bridge or a ford close to the mill and all this could lead to the establishment of a small settlement. Variations of Milton (mill-town) are common place names all over Britain but in Eastern Scotland the village name is sometimes added, as in Milton of Kildrummie (Aberdeenshire) or Milton of Balgonie in Fife (cf. *Coaltown* of Balgonie nearby, *Kirkton* of Glenisla and *Fishertown* of Ussan, by Montrose). No such names occur along the Water of Leith but several settlements originated around mills, including Stenhouse and Bell's Mills (both possibly 12th century), Newmills downstream of Balerno, and only "New" in 1604 and, in the 19th century, Malleny Mills on the Bavelaw Burn, now part of Balerno.

Of course these were never towns in the modern sense, but like the ferm-touns (farm towns) simply clusters of assorted buildings with homes for a few families, making up a hamlet in an outlying part of the town or village at the centre of the parish. The mill settlements would be linked by a track along the riverside and odd fragments of such a roughly-paved way can be seen in Colinton Dell between Bog's and Kate's Mill sites.

CHAPTER 3. The surviving traces.

The aim of this chapter is to survey the bits and pieces of water-powered industry that were present on the Water of Leith, to say how they compare with those elsewhere, and to define what now remains for the interested observer to see. It will be easiest to start at the upstream end of our hypothetical mill layout and work downstream.

Weirs

Some weirs (old local name damheads) were demolished or breached when their mill went out of use, to stop the nuisance of water entering the lade, and at least two weirs (for Westmills above Colinton and the Juniper Green weir serving the two Woodhall Bank mills on the north bank) have been removed completely in the last 20 years. Other weirs have been restored to keep them safe, like Redbraes (Bonnington). The river was not powerful enough to drive a mill on either bank, so there are no weirs with matching lades at either end (cf. Kinnaird on the South Esk at Brechin or Upper Blantyre on the Clyde) and the weirs are set obliquely towards a lade on one or other bank. The two Dean weirs are probably the most spectacular structures, at about 12 feet high, but the long V-shaped apron weir in Colinton Dell supplying Redhall Mill is impressive, and - with extra water in the river - Kinleith (Currie), Mossy, Gorgie, Bell's and Redbraes (Bonnington) can certainly look the part. Slateford weir is currently holding out in a state of disrepair and the western end has been progressively covered up since the 1967 Lanark Road bridge replacement. A shallow weir just upstream of the bridge may mark the site of the ford that served as the original river crossing referred to in the place-name Slateford but seems to have no other significance. The small (1-2 foot) weir just above Balerno Bridge is purely ornamental and two small weirs upstream of here (at Larchgrove and Glenpark) appear to be relatively modern structures, perhaps for small generators.

Mill ponds and reservoirs

Ponds were rare on the river. Canonmills had its loch to store water and Leith Sawmills may have had a pond, but these are the

The sluice at Woodhall Board Mill, with three gates set in a brick box structure, presumably designed to trap debris before the water passed into the mill via a tunnel.

exceptions. Of course, from the mid-19th century Harperrig, Harlaw and Threipmuir reservoirs, with a total area of 485 acres (196 hectares) and a capacity of 1,579 million gallons (7,106 million litres), must have made summer working less of a gamble for the millers and restricted the force of winter floods that had previously caused great damage to the mills. In 1659, for example, "11 mills belonging to Edinburgh and five belonging to Heriot's Hospital, all on the Water of Leith, were destroyed on this occasion, with their dams, water-gangs, timber-graith and haill other warks." (Cumberland Hill). In 1739 Helen Souter, a wet-nurse, was carried away at Royal Mill and her body was swept over 15 dams, Currie's minister the Rev. John Spark and Alexander Bowman of East Mill being drowned in the same flood (John Tweedie). In October 1832 there were three days of flooding in which 12 weirs were damaged and Slateford Bridge was brought down (Gladstone-Millar).

Today the reservoirs in the Pentland Hills are used for recrea-

The weir and sluice for Balerno Bank on the Bavelaw Burn

tion, notably trout fishing, add priceless landscape value to the views of the hills from the north and still retain their role in flood control. Even now, however, the river can still show its power. On April 25 and 26, 2000, 48 hours of continuous rain, the amount expected in a month, brought the river level almost up to the underside of Slateford Bridge, flattened the allotments at Chesser, flooded Murrayfield Rugby ground and forced residents from their homes at Roseburn, Canonmills and Warriston. There was more minor damage to walls and fences and the appearance of the river was grossly altered, with deep banks of stones and boulders where there had been deep pools.

Sluice-gates

Being made of wood and standing in water, sluice-gates are seldom long-lived structures and specimens in reasonable health 25 years ago (Kinleith and Redhall) have now succumbed to rot and petty vandalism. The best examples are, not surprisingly, the most modern and can be seen on the main river at Woodhall just beside the weir, where the sluice is a complicated brick and metal structure with three gates, and on the Bavelaw Burn, Balerno, along the lane to the Kestrel Hotel, where the lade served Balerno Bank Paper Mill with water for processing into the 1970s.

Lades

Other examples of town lades can be seen at Haddington (River Tyne), Blairgowrie (River Ericht) and Perth, where water is drawn from the River Almond and travels 4 miles into and across the city, once driving a series of bleachfields, a foundry, a waulkmill and then the City (grain) Mills complex before entering the River Tay. Greenock boasted the Shaws Water Aqueduct (1827), fed from a massive artificial lake, Loch Thom, named after its creator Robert Thom. The aqueduct ran 5½ miles to the head of 18 falls which powered a whole range of industries.

Edinburgh's Great Lade, drawn from the river at Dean and travelling for almost 2 miles across what are now the inner suburbs, stands comparison with Perth's for length and activity and was described by Cadell in his essay in *The Water of Leith* as "one of the civil engineering wonders of the city." He traces the route of the lade beyond Greenland Mill in a wooden trough set on stilts, under St Bernard's Bridge, along the back of the flats in Saunders Street and across Kerr Street to Stockbridge Mill. It crossed under Clarence Street into Silvermills and turned north down East Silvermills Lane before passing behind the houses on the south side of Henderson Row. It then crossed Dundas Street and followed Eyre Place to Canon Street and the Canon Mill. From the bottom of Canon Street it went behind houses in Monro Place and entered Beaverhall and Logie Green, where it is difficult to relate the route to present street names, before returning to the river by St Mark's Bridge.

Other lades on the Leith were, of course, shorter but that at Dean on the north bank drove the Sclait Mill, the West Mills and several others (of which, sadly, we have no details) and appears on a photograph by the Aberdeen photographer George Washington Wilson, while the lade at Slateford drove at least three mills in succession. The lades at East Mill (Currie), and Juniper Green

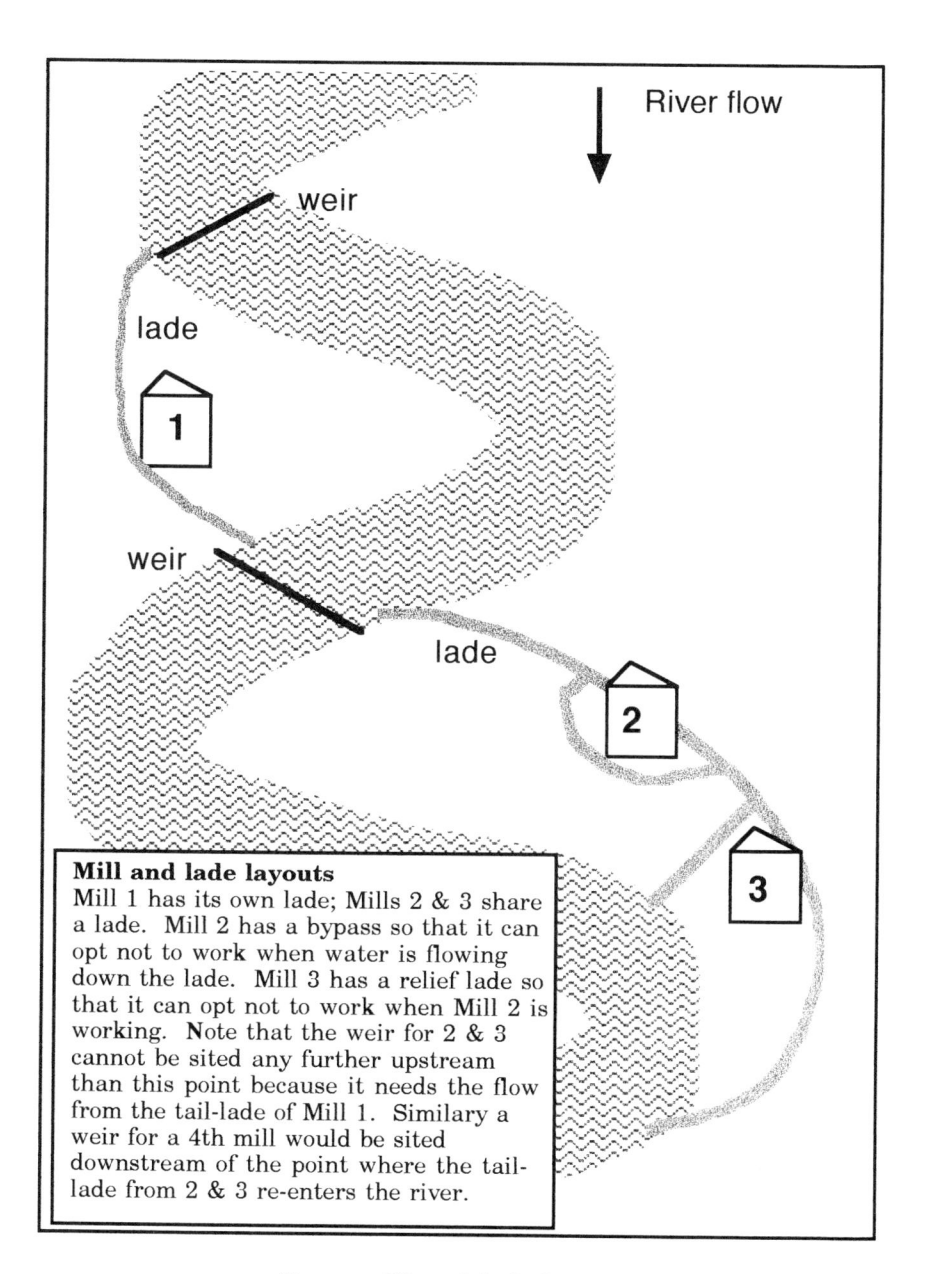

Mill and lade layouts
Mill 1 has its own lade; Mills 2 & 3 share a lade. Mill 2 has a bypass so that it can opt not to work when water is flowing down the lade. Mill 3 has a relief lade so that it can opt not to work when Mill 2 is working. Note that the weir for 2 & 3 cannot be sited any further upstream than this point because it needs the flow from the tail-lade of Mill 1. Similary a weir for a 4th mill would be sited downstream of the point where the tail-lade from 2 & 3 re-enters the river.

Some mill and lade layouts

each drove a grain mill and then a snuff mill. The Redhall (Colinton) lade was powerful enough to drive Redhall and then Kate's papermill and gurgles along merrily to this day, even driving a one year-old wheel. The Gorgie lade was a long one, driving the Gorgie Mills complex and then continuing to Murrayfield (Dalry); traces of its route can still be seen on either side of Stevenson Road. Finally, the lade at Bonnington drove mills on both sides of Newhaven Road. There is nothing here to rival the massive cotton mill lades at New Lanark on the Clyde and Stanley Mills on the Tay, where the water travels through a tunnel, but those mills were set in relatively open country where long lades could be constructed: the density of mills in the narrow valley of the Water of Leith did not allow this.

A casual look at the river can suggest that mills have just developed haphazardly along the banks, here close together (at Dean, Colinton and Juniper Green, for example), there farther apart. In fact the siting would be influenced by several factors, including access, topography, land ownership and, above all, the presence of other mills. Old photographs of Currie below Mutter's Bridge and the river at St Bernard's Well (page 65) show one of the difficulties - the riverbed carries only a trickle of water because most of the flow was out of sight travelling down the lade. There would be no prospect of inserting another mill on the riverside at these points and it is no coincidence that the boom in new mill development around 1800 occurred well upstream, at Balerno, where such restrictions of space did not apply.

Given a suitable site, the positions of the weir and the mill along the lade were also critical. The weir needed to be as far upstream as possible to give the maximum head of water, yet still be below the outflow of the next mill if it was to collect enough water (see page 17). A long lade from weir to mill would increase the head and hence the power for a larger wheel. Another option was to place the wheel in a deep pit, so that a larger (taller) wheel could be accommodated. This however would require a longer tail-lade because the water leaving the wheel would be below river level and the lade could only rejoin the river at a point where

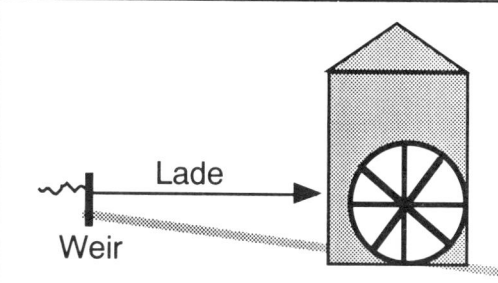

Short tail-lades: water leaves the wheel at or above river level and can return almost directly to the river without any risk of back flow and flooding the wheel. All the mills around Colinton were of this type.

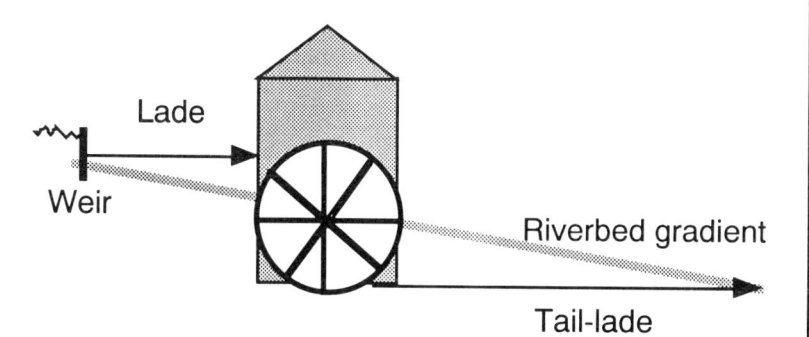

Long tail-lades: With larger wheels sunk into deep pits, the bottom of the wheel could be below river level with the danger of a back flow of river water, stopping the wheel. A long tail-lade would carry the water away from the wheel (by sloping slightly) and return it to the river at a point downstream where river water would not enter the lade. Examples are the lades at Newmills and Kinauld.

Explanation of long and short tail-lades

the water would not flow back up the lade and drown the wheel. Long, deep lades (100 yards) were used at Newmills, where the wheels may have been overshot, and at Kinauld. Some other mills had literally no tail-lade at all (Mossy, Upper Spylaw, and the West Mills at Dean).

Sharing of lades demanded a degree of cooperation from the millers: if the main sluice at the weir was closed no-one could work. If it was open, the mills needed a bypass arrangement for their wheel(s) if they were not to work: to simply close a sluice onto the wheel would stop the wheel but cause the lade to flood. Traces of such bypasses, or relief sluices, can be seen on the lades at East Bank Snuff Mill and at Watt's Snuff Mill (Juniper Green) and a line of trees across the middle of Roseburn Park marks the route of a relief lade that turned the water back to the river short of the wheels at Dalry Mill.

Water wheels

Henry Dempster's survey of the mills on the river seems to have been undertaken as a preliminary to the building of the reservoirs. Unfortunately its precise date is uncertain within the period 1842-54 but in arguing for the construction of a new reservoir at Buteland Hill it seems to pre-date Harperrig Reservoir (1856) west and upstream of there; the date 1850 has been used throughout this account. Like most of the information about the mills of the upper valley, we owe the discovery of the document to Currie's local historian, John Tweedie.

Dempster lists 37 mills then operating on the main river with the dimensions of the wheels (diameter or height x breadth), the height of the fall from the weir and the use being made of them (paper, corn, barley, snuff etc.,). Although a power rating in horse-power (h.p.) is given for each wheel, no precise calculation from the flow rate, the wheel size and efficiency was attempted. Instead, the figures are arbitrarily based on the precept that 8 horse-power is needed to drive one pair of stones at full speed, with 12 h.p for 2 pairs and 16 h.p. for 3 pairs. Only grinding mills had pairs of stones, so the horsepower of wheels in other types of mills must have been assigned on an even more arbitrary basis and the 10 x 10 feet float wheels at Gorgie, presumably situated in the lade itself because the fall is given as 0 feet, are rated at 2 and no horse-power at all. Nevertheless, the data are an invaluable snapshot of the water-powered industry and its water needs, the 37 mills using 66 wheels with a total of 828 h.p.

The largest wheels are those at the papermills: Kinleith's wheel (23 x 10 feet) was 36 h.p., as was Balerno's 18 x 12 feet wheel. Next come the 26 h.p. wheels at Mossy Mill (18 x 6 feet), Kate's Mill (14 x 12 feet), also papermills, 26 h.p. wheels at the largest of the grain mills, Bell's (16 x 6 and 16 x 8 feet) and Water of Leith (West) Mills at the Dean (14 x 12 and 1 8 x 9 feet). At the other extreme are 8 h.p. wheels like that at Upper Spylaw Snuff Mill (16 x 3 feet 8 inches) and a 6 h.p. wheel (14 x 4 feet) at Logie's (stone-cutting) Mill.

The largest wheels mentioned in the survey were obviously sufficient to power a range of machinery in a small factory, but steam engines were probably already helping out where the demands of the machinery or low water levels made it necessary and many local farms already had their own steam engines for threshing. In more national terms, none of these wheels is particularly large. The most spectacular, and most visited, British water wheel is the Lady Isabella Wheel at Laxey, Isle of Man, which is 72 feet high by 6 feet wide and developed 185 h.p. at 2½ revolutions per minute, pumping 250 gallons of water per minute from the lead mines. In Scotland, Sir William Fairburn built wheels for the textile mills at Catrine in Ayrshire which were 50 feet in diameter and 10 feet wide and developed 250 h.p. and the cotton mill at Greenock had what was for a time the largest wheel in the world, 70 feet high and capable of 200 h.p. Such vanished wonders aside, wheels capable of well over 100 h.p. exist on the River Ericht at Blairgowrie and were still being installed there in flax and jute spinning mills around the time Dempster's survey was made.

Most wheels on the Water of Leith were undershot or breastshot, the exceptions being on the Bavelaw Burn, where there was a fall of 300 feet between Harlaw and the junction with the main river, a distance of only 1.7 miles and a nominal flow rate of 300 gallons per second after enlargement of Threipmuir Reservoir in 1889. Compare that with the main river, usually regarded as swift, where the overall fall is almost 1,000 feet from Harperrig to Leith (about 20 miles). The Bavelaw Burn was thus ideally suited to large diameter, overshot or pitchback wheels, and those at Balleny, The Glen, Balerno Bank and Byrnie's were of this pattern. We have no details of the Spinning/Bung Mill. The remains of wheels at Balleny and The Glen survive but are not on public view. On the main river the only mill wheel on display is the restored undershot wheel at Bonnington (see page 71). This is about 25 feet in diameter and has simple wooden paddles arranged radially. It is an impressive sight but would be far more exciting with a water supply and some movement.

Mill machinery

If there were, as one claim has it, once more than 50 grain mills on the river among the total of 76, with at least one per settlement, it is amazing that we have only the group of three stones at Lindsay's Mill in Dean Village (page 24) on public view. Millstones would constantly wear thin and be replaced and others would be left in place when mills finally closed down. In other parts of Britain, millstones are prized as paving stones or are preserved and set on end to hold a sign with the name of the village. Here they have almost all vanished - no small curiosity when a stone was commonly four feet in diameter and would weigh up to a ton: it could hardly have been spirited away in a coat pocket or even a wheel barrow. Perhaps the stones were broken up for use in walls or garden rockeries - we simply don't know. One hint from the 1897 lease of Watt's Snuff Mill (see page 53) is that while the mill, wheel and pit wheel belonged to the laird, the machinery inside, including the stones in a grain or snuff mill, could belong to the tenant and be his to dispose of, perhaps by selling on to another mill, if he went out of business.

Soft grains, like oats and barley, and peas and beans, could be milled using gritstones of the type produced in Derbyshire and Yorkshire (hence "millstone grit" for the rock) but similar stones

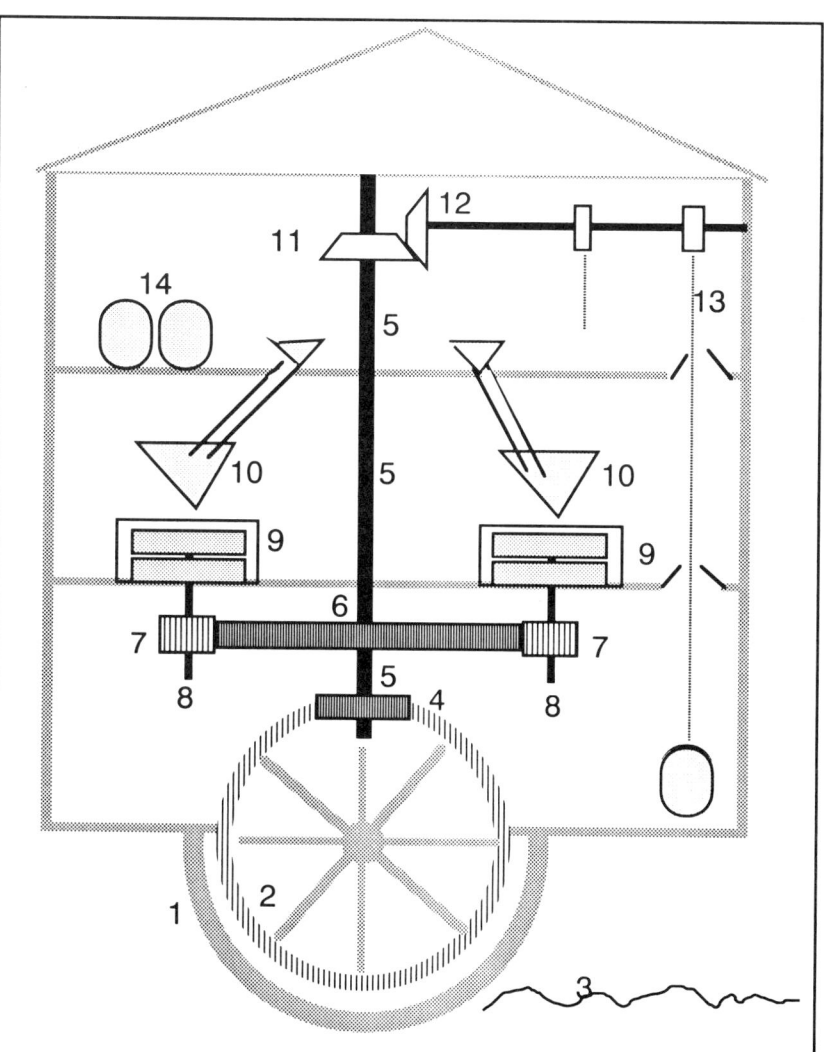

Layout of a three-storey grain mill, much simplified.
1. Waterwheel, outside the building. 2. Pit wheel, inside, on same axle. 3. Tail lade. 4. Trundle or wallower (main horizontal drive) . 5. Main drive shaft. 6. Spur wheel. 7. Stone nuts or pinions. 8. Secondary shafts driving runner (upper) stones. 9. Pairs of horizontal stones in wooden casing. 10. Hoppers to feed grain into the stones. 11. Crown wheel. 12. Drive shaft for sack hoist and accessories. 13. Sack hoist with trapdoors. 14. Storage for grain sacks. There would be grain elevators if meal from one pair of stones was then re-ground in the second.

French burr stones at Dean Village. The nearest one displays traces of a spiral dressing

were also produced locally at Hailes Quarry (1750-1900?) just north of the Lanark Road and the Union Canal. Wheat required a harder, finer stone and the Dean stones are examples of French burr stones, composite quartz stones imported from Normandy. Pieces of stone were shaped to fit together, cemented with plaster of Paris, and bound together with an iron rim. Such stones were twice as expensive as gritstones. An advert for J Smith & Sons, of the Scottish Wirework and Millstone Manufactury, 219 High Street, Edinburgh, describes them as importers of French burr stones. The dressings, the distinctive pattern of grooves cut into the flat surface of the stones, are still just discernible on the stones at Dean. Oddly, the stone facing the river seems to show traces of an unusual spiral dressing while the other two have the common dressing of parallel grooves in tangential segments. These stones are said to have come from Lindsay's Mill, but Lawrence Walker, formerly of Bell's Mills, regards the West Mills as more likely.

Millbank House, associated with Bog's Mill

A few stones survive at Bell's Mills. One is incorporated into the fountain at the granary and others currently lie close to the gateway of the mill house. These are one-piece stones of different sizes. The smallest at 2 feet diameter displays the common dressing of parallel grooves and was used for grinding spices. Two stones located in gardens in Juniper Green and in Currie are one-piece stones with no sign of a dressing. These stones are probably of the edge-running type, standing vertically and rotating about a horizontal axis. The Currie stone, of grey gritstone, is 43 inches wide and 4 inches thick at the edge but nearly 8 inches thick at the octagonal eye. Lawrence Walker identified this as a snuff stone and it probably came from the East Mill site close to where it now lies. The Juniper Green stone is three feet in diameter, of reddish colour and uniform thickness with rounded edges. From my photographs, Lawrence Walker identified this as a barley stone, and the site suggests it may have come from Woodhall Bank Barley Mill. A broken stone, originally 4 feet in diameter, lies in the river upstream of Slateford.

The former Baxters' granary at Bell's Mills

Cast iron spur wheels or crown wheels lie beside the pathway at Redhall Mill but elsewhere everything has disappeared, perhaps in the collection of scrap iron for the war effort in 1940-43, so that we have no relics of snuff-milling even from fairly recent time, let alone a set of fulling stocks from the more distant past.

Mill buildings: mills, granaries, kilns and houses

As in other parts of Britain, the mills that have survived best are the small scale enterprises that did not develop into major commercial concerns (and which were lucky enough not to burn down). The papermills, for example, were continually altered and expanded in recent times on large sites that were valuable for residential or other industrial development, but the grain mills were often of a scale suitable for conversion into one or more private homes. So we have fine several-storey mills of elegant stone at Spylaw and Upper Spylaw, at the Dean's West Mills and at Canonmills (now offices). The massive granary at the Dean (the Baxters' Tolbooth, close to the southern end of the old bridge)

The Baxters' stone taken from the doorway of their Jericho granary, displaying the all-important sun, cherubs, a set of scales, sheaves of corn, an hour glass and peels for loading dough into the ovens. The inscription from Genesis reads "In the sweat of thy face shall thou eat bread. Anno Dom. 1619." The stone is now set into the wall of the former inn, The Baxters' House of Call, at the end of Dean Bridge.

is now flats, its age somewhat disguised by alterations and the harling over its stonework, but given away by the small half-shuttered windows surviving on the stair towers. It still has its Baxters' insignia, the wheatsheaf and crossed peels (shovels used to load the dough and finished bread into and out of the ovens), scales, cherubs' faces and the legends "GOD BLESS THE BAXTERS OF EDINBRUGH UHO BUILT THIS HOUS 1675" and "God's Providence is our Inheritens". Jericho, another tall granary, or girnal, owned by the city of Edinburgh, was built in 1619 on the site of the architect's offices in Miller Row. It burned down in 1956 but its superb datestone can be seen on the wall of the house at the south end of Dean Bridge, complete with Baxters' symbols and the verse from Genesis 3: 19 *"In the sweat of thy face shalt thou eat bread."* The granary at Bell's Mills (Hilton Hotel) also bears the Baxters' wheatsheaf with the date 1807.

The kilns that were once stood beside or formed an integral part of many grain mills have disappeared except at Currie, beside Currie Brig (page 48), and at Royal Mill, hidden away on the river above Ravelrig House at Balerno. Both are circular, free-standing structures. Mansion houses for the owners exist at Redhall (George Inglis, 1750), Stenhouse (enlarged by Patrick Eleis, 1623, see page 62), at Spylaw Park where James Gillespie's mansion (1779) fronts the former snuff mill (page 54), at Murrayfield (Mungo Russell's Roseburn House, 1582) and at Currie (former Glenburn Hotel above Blinkbonny, 1896) where the Bruces of Kinleith Mill added an observatory to their home. Fine millers' houses can be seen at Bog's Mill (Millbank, pre-1735; see page 25), Newmills (469 Lanark Road West), Bell's Mills, beside the Brae (see page 38) and Bonnington (Bonnyhaugh House, 1621).

CHAPTER 4.
The Water-powered Industries of the Water of Leith

Before we consider the many industrial uses of the river in more detail, it is worth noting some industries that were not represented here. Although Scotland's first cotton mill opened in 1774 in Penicuik, just across the Pentland Hills from the Water of Leith, the more significant developments were elsewhere. Cotton manufacture came to New Lanark on the Clyde in 1785 and the site became the world's largest water-powered cotton mill under the enlightened leadership of Robert Owen. Glasgow became a cotton port and a major industry developed in its hinterland. In 1700 Leith was Scotland's first port but was overtaken by Glasgow and Greenock in the next 100 years. Perhaps Edinburgh and the Water of Leith were on the wrong coast and too distant to take part in the cotton boom (although Newmills barley is said to have been shipped from Glasgow to the West Indian slave plantations). Even the early cotton mills were big and demanding in terms of the water power available, with sites on the major rivers Tay, Don and Clyde. Cotton manufacture was never attempted on the Water of Leith.

No significant woollen manufacture developed either, despite waulking of hand-loom products from an early time. Rope-making flourished in Leith, second only to seafaring in terms of employment by the mid-18th century, but occupied rope walks in North Leith and by the Links, well away from the river and its water power. Although many small foundries developed latterly in Edinburgh, including Mathers of Fountainbridge, who provided a replacement water wheel for East Lothian's Preston Mill early in the 20th century, none were water-powered as far as we know and even the metal-working associations of Silvermills are uncertain. There was a brief and small-scale attempt to mine copper at Currie in 1758. Perhaps the massive Carron Ironworks at Falkirk, equipped with a range of water wheels by Smeaton in 1764-85 and forging cannon for the Napoleonic Wars, cast too long a shadow. Gunpowder manufacture flourished around Edinburgh, at least on the Linhouse Water at Camilty 10 miles west of Balerno, and on the River Esk at Roslin in Midlothian. On the Water of Leith the name Powderhall is thought to commemorate a similar enterprise in 1695 but we don't know whether it was water-powered and the site is now associated more with athletics, greyhound racing and, more mundanely, waste disposal.

Grain milling

Even without the power and influence of the Baxters at their headquarters in Dean Village, the milling of grain would still be central to the river's use by virtue of the mills regularly spaced along its course. Many millers enjoyed a feudal obligation for the local inhabitants or tenants to bring their grain to the mill and pay for the milling with a fraction (classically one-sixteenth) of the product. They were "thirled" to the mill and their payment to the miller was termed multures.

The Baxters were just one of the Incorporated Trades of Edinburgh, amalgamated as long ago as the 15th century. The date of the Baxters foundation is uncertain but they were confirmed in 1522. The various crafts and guilds acted like trade unions, friendly societies and insurance companies, with their own structure, discipline and even a particular alter in St Giles' Cathedral to focus worship by the craft (and maintain their status?). Other trade guilds included the Hammermen (wrights, smiths, saddlers, metal-workers, incorporated 1483), Wrights and Masons (1475), Websters (weavers, 1475), Fleshers (1488), Waulkers (1500), Candlemakers (pre-1582), Barbers (originally combined with Chirurgeons, 1505), Tailors and Bonnet-makers (1530) and Skinners and Furriers (1450). The guilds drew up rules for apprenticeships, with appropriate tests to be passed.

The Baxters held gala days, including the annual Feeing of the Millers, when a stately procession of officials visited the various mills at the Dean, agreeing wages and other business before settling down to the merrier business of food and drink and returning much later in an altogether less stately fashion.

The miller's was a skilled craft, with constant vigilance for watercourses, the sluices, the water wheels and the millstones.

Water levels had to be watched continuously and the stones had to checked against the dangers of running dry (of grain) which brought the risk of damage and even sparks and fire - a calamity that saw the end of many mills on the river. Traditionally the miller was not popular with all his customers, but about 1801 Currie's poet, James Thomson, sent his good wishes to the miller at Newmills, John Stein, in the following verses:

Lang may your mill keep haill an' weel,
May naething skaith her outer wheel
Nae bits of flint, nor chips o'steel, gang through the happer
To spoil the stanes, an' gar them reel an' stop the clapper.

In summer drouth, when dams rin dry,
May springs o' water yours supply:
To stop ilk hole an' bore ay try in your dam-head,
That naething may rin through or by but down the lead.

Health to your mill this mony a year,
An' may she ay hae rowth o' beer,
To had the barley-stanes asteer for what is best
An' when your purse rins oer with gear, een tak your rest.

Before the 19th century the grains would be oats and barley; peas and beans would also be ground. The barley was a primitive variety, called bere (confusingly spelled *beer* in the last verse of Thomson's poem), with relatively few grains per stem, and recalled by the farm name Boll (weight of a standard measure or sack) of Bere, along the Lanark Road from Balerno. Grain was also a form of currency: farm and mill workers might be paid in part with a boll of oatmeal (about 120 lb). Wheat came to Scotland later and the availability of harder North American wheats signalled the end of the country corn mill. Steam-powered roller mills appeared close to the point of supply (Leith) and the small antiquated water mills tucked away in their inaccessible sites upstream could not hope to compete. The huge Chancelot Mills appeared at Bonnington, to look down on Bonnington Mill and watch its slow decline grinding grain for animal feed, while still bigger mills shot up almost on the quayside at Leith for the national milling concerns like the Rank Group.

Waulking and textiles

Coarse woollen and possibly linen cloths would have been produced on spinning wheels and handlooms in cottages long before our first records of the Water of Leith mills and would have been softened by soaking, rubbing and beating by hand and foot prior to the use of water power for the purpose. Rhythmic waulking songs that lightened the task are preserved in the folk music of the Hebrides and may have had their Lowland equivalents. We have the surname Walker (Waulker) as a reminder of this trade, along with the English equivalent Fuller (hence fuller's earth, a clay-like paste used to soften the cloth). Latterly, Walker was one of the most common names among the millers of the Water of Leith.

There are records of weavers at Picardy, off Leith Walk; at Bonnington, where Flemish workers were introduced as teachers in the early 17th century; at Slateford; at Stenhouse; at Dean Village; and fainter traces in the street name Weaver's Knowe at Currie and the occupation of the local poet James Thomson, so this was probably a universal activity, with weavers in every parish. The regular occurrence of waulk mill sites along the river, for example at Currie (Balernoch, Kinleith), Colinton (Mossy, West Mills, Hole, Bog's), Slateford, Stenhouse, and so on, certainly supports this idea. Some of the sites gave rise to, or were combined with, bleachfields, notably at Slateford, where in 1822 the Union Canal Company were obliged to pay compensation because their new aqueduct overshadowed and thereby denied vital sunlight to the bleachfield. Many bleachfields subsequently developed into dyeing and laundering businesses, including A & J McNab's Inglis Green Laundry at Slateford. In fact the word laundry may come from one name for the wooden trough (launder) bringing water to an overshot wheel.

Flax was grown locally, for example around Colinton, and

Woodhall was recorded as a lint (flax) mill in 1779 and Kirkland in 1777. The Edinburgh & Leith Rope Company's Malleny Mills (1805) used flax and hemp to make sailcloth. If the raw flax was processed and spun there, as the presence of retting ponds and a spinning mill confirms, and the cloth was bleached and dried on the green before being smoothed and polished at the beetling mill (now The Glen), where was it woven? Was it taken all the way back to Leith and then returned for finishing or was this weaving also a local (handloom) activity? One hint at the first alternative is that there were two Rope Companies in Leith in 1778, with a contemporary claim that "together they employ 120 looms for making canvas..." (Marshall). This suggests that even this relatively small enterprise on the Bavelaw Burn gave plenty of work for the local carters with raw materials, yarn and finished cloth passing backwards and forwards between Leith and Malleny. They may well have complained about the congestion and surfacing on the Lanark Road, as those who live around it do today.

Paper-making

The basics of paper-making are simple. A source of cellulose fibres (linen or cotton rags, grass or wood pulp) is chopped up and broken down in water so that the fibres are brought into suspension. The suspension is caught on a sieve, forming a web of uniform thickness, which is then squeezed and dried as separate sheets, or in later days as a continuous roll. The spread of paper-making technology was, however, agonisingly slow. Paper mills existed in France in 1338, in Germany in 1390, but not in England until 1495 and the craft came to Scotland in 1595. From 1711 paper was taxed and papermills had to be licensed, so that there are official records of the Scottish mills with their outputs and these have been diligently researched and reported. Waterston's study of the first (pre-1700) papermills around Edinburgh is particularly relevant to the Water of Leith. AG Thomson's more exhaustive account *The Paper Industry in Scotland 1590-1861* goes over the same ground and reproduces lists of Scottish mills for 1825, 1832, 1852 and 1853.

These writers make it clear that the Water of Leith had a

central role in the early days of Scottish paper-making, beginning in 1590 when Mungo Russell and his son Gideon of Dalry Mills contracted with two German paper-makers, Peter Groot Haere and Michaell Keysar, to convert one of their mills to make paper. Mungo Russell died in 1591 but the venture is thought to have lasted for 15 years. The next four Scottish mills were also on the Water of Leith, being at Canonmills (1652), Dalry again (1674), Upper Spylaw (1681) and Canonmills again (1682). Thereafter mills appeared at Woodsyde, Glasgow (1683), Restalrig, Edinburgh (1686), Cathcart, Glasgow (1686), Ayton, Berwickshire (1693), Braid, Edinburgh (1695), Yester, East Lothian (1695) and Aberdeen (1696). Thomson estimates the total production by 1700 to be 150-200 reams per week and that home production then matched the amount of imported paper.

One of the early paper-makers, the engineer Peter Bruce, sometimes called de Breusch, is especially interesting. An imigrant from Flanders, but referred to as German, he was involved in a range of projects between 1674 and 1690 and was twice imprisoned, some of the animosity towards him deriving from his Roman Catholicism. His first projects were a piped water supply from Comiston to Edinburgh (1674), a harbour at Cockenzie (1678), a water pump for emptying mines and an engine for cutting iron (both patented in 1680), and a papermill at Canonmills (1681) with a patent for making and selling playing cards (1682).

The papermills - there may have been two - brought trouble with the principal tenant John Patterson and a local farmer called Alexander Hunter. Hunter was found guilty of riot and fined £50 following a disturbance during which Bruce's wife Louise was thrown into the dam (the lade? Canonmills Loch?). Bruce moved to Woodsyde, Glasgow, in 1683 and set up paper-making there, visiting Ayr to raise a sunken ship, but trouble was again waiting and although he won a case against his Woodsyde partners he was imprisoned for 8 days. He returned east to set up a papermill at Restalrig and became the King's Printer in 1687 after successfully petitioning for an embargo on imports of certain foreign papers. He was again imprisoned briefly and if religious differences and xenophobia may have caused his many troubles, resentment and jealousy occasioned by his obvious ingenuity and enterprise may also have played a part.

In the 18th and 19th centuries paper-making continued and prospered on the Water of Leith, with major developments at Kate's Mill, the West Mills at Colinton, Kinleith and Balerno (Kinauld) but there was increasing competition from mills on the River Esk in Midlothian and indeed the Cowan family, once at Kate's, became one of the driving forces in paper-making on the Esk at Penicuik. The big event in the 19th century, after the arrival of the Union Canal in 1822, with only marginal influence to the mills on the river, was the creation of the Balerno Branch of the Caledonian Railway, opening eventually in 1874. The story is well told by Donald Shaw in his book *The Balerno Branch and the Caley in Edinburgh* and the following summary is drawn from his account.

By 1864 there were seven paper mills and the main argument for the Balerno Branch was their demand for raw materials, like coal, rags and grass, and delivery of their product. Ironically it was the mill owners, who were also the principal landowners, who were at first the main opposition. They voiced their complaints, on grounds of loss of amenity, nuisance in the building of the railway (the immorality and disorderliness of the navvies), and the general unpleasantness of paper-making and railway locomotives (smoke, noise), at a meeting in 1864 but by the following year the Bill to authorise the branch had been passed by Parliament. There were still several hiccups and many problems concerned with engineering a line up the steep, narrow, twisting valley, the gradient reaching 1 in 50 in Colinton Dell. Part of the solution was the building of four special short wheel-based bogie locomotives (Twelve more 0-4-4 tank engines, known as "Balerno Pugs," were built later, in 1899). Other preparations included purchase of the mills of Hailes and Newmills, the engaging of William Arrol, who later designed the Forth Bridge, for the ironwork of the many bridges and various payments of compensation, including £1450 to the tenant of Redhall quarry and £4,000 to Sir James Liston Foulis of Woodhall (Juniper Green). The final cost of the branch line was £134,000.

The success of the Caledonian main line railway to Glasgow (1841) was immediate, and caused the rapid decline of traffic on the Union Canal. On the Balerno Branch, sidings soon opened at Kinleith (1875), Kate's Mill (1879), Colinton Mill (1889) and Balerno (Kinauld glue works) in 1904. Kate's Mill siding closed in 1890 when the mill burned down but was used later to take woodflour produced at Redhall. The accompanying passenger service aided the development of Colinton and Juniper Green as commuter suburbs for Edinburgh with a spread of houses for the wealthy middle-class. There were only four services each way every day in 1875 but 20 by the 1930s. Yet in 1883, with bookings for 117,133 passengers, income from goods still outweighed passenger income more than three-fold.

Paper-making survived into the second half of the 20th century at Mossy Mill, Woodhall, Kinleith and Balerno Bank, though use of water for power was occasional (at Kinleith) at most. John Tweedie tells us that Kinleith's most prized product, "Featherweight", a light but bulky paper used for printing novels, "was the backbone of the industry for 70 years " and that Kinleith-trained papermakers pioneered similar industries in other parts of the world. Sir David Henry left Juniper Green at 14 and eventually became chairman of New Zealand Forest Products Limited in Tokoroa, where the company's mills were named Kinleith. James, Edward and William Findlay also left Juniper Green, in their case for Canada and the USA, and erected mills in Toronto and St. Catherine's, Ontario (the latter also a Kinleith). Despite this powerful influence, the Water of Leith's mills closed in the 1960s or 1970s, with Balerno Bank lingering on as a plant for coating paper made elsewhere once its own paper-making days were over. It was demolished in 1989 and Balerno's major source of local employment had disappeared.

Snuff milling

Some mills were dedicated to snuff-making while others,

usually the grain mills which could change products quickly, made snuff for short periods. Thus the snuff mills at Colinton (Gillespie's at Spylaw), Currie (East Mill) and Juniper Green (Watt's) never seem to have made anything else, but Upper Spylaw, Bog's, Dalry, Slateford, Stenhouse and Richardson's (Kirkland) all had spells making snuff.

The use of snuff - the sniffing of a pinch of finely-ground tobacco stalks and leaves - seems to have begun in the 18th century and still lingers on today, though silver snuff boxes have become collectors' items. James Gillespie's prosperity arose from his hard work and canny business sense: his brother John ran the shop in Edinburgh's High Street, while James ran the mill in Colinton, and remained a bachelor, saving his money for the long-term benefit of Colinton and Edinburgh.

An engraved notice issued by Richardson Brothers about 1850 proudly proclaimed them as "Snuff Manufacturers in Scotland to Her Majesty Queen Victoria and His Highness Prince Albert" with text as follows:

TO SNUFFERS

The sale of nearly Four Hundred Tons of Richardson's Rappee within a few years has induced us to build a new Snuff Mill at Spylaw, Colinton, combining the most improved methods of Snuff grinding, both on the Scotch and English systems. Our Rappee is manufactured from a selection of Tobacco, grown in every quarter of the world, and will be found an agreeable, pungent pinch.

The notice was presumably issued to coincide with development of Kirkland Mill (formerly a meal mill which they re-named Spylaw) following their move downstream from Gillespie's mill at Spylaw a few years earlier. The illustration that heads the notice shows pairs of edge-running (vertical) stones grinding in shallow dishes and powered by an overhead horizontal shaft. A very similar arrangement of vertical stones survives for chalk grinding at Thwaites Mill near Leeds and can be seen in operation.

Despite the proud claims of the Richardsons, this was a small-

Lawrence Walker and the mill house at Bell's Mills in 2000

scale industry, well designed to last. The product was relatively small in volume but high in price (compared to paper or grain) and a single worker could run a mill with a small wheel to drive the machinery and perhaps a single pair of stones, or their equivalent, and thus milling went on at Currie until 1920 and at Juniper Green until 1940, with the Watt family latterly involved at both mills.

Timber and wood-working

There were water-powered sawmills at Leith, two sites on the Bavelaw Burn at Balerno (one a former flax mill converted to a bung mill) and the Cockburn estate sawmill on the Cock Burn. It is perhaps surprising that there were not more sawmills but the last two hundred years did see the development of another wood-based industry, the grinding of sawdust to make wood flour, which became special to the Water of Leith. This was almost entirely the province of a single family, the Walkers.

In 1880 Gideon Walker occupied Greenland Mill, just below Dean Bridge, using it to produce provender (feed for animals, principally horses) by day and grinding wood flour at night using the same sets of stones. When the Great Lade was closed in 1890 Greenland was literally left high and dry and the Walkers moved upstream to Bell's Mills. The wood flour was sent to Fife for the linoleum industry and a very fine powder was needed, so that 100 mesh screens were employed, with 100 wires to the inch and horizontal mill stones were coated with emery using a magnesium cement. For certain inlaid linoleums the wood flour product soon replaced cork.

From 1900 Gideon's son Allan was also involved in a fast-developing enterprise and leases on Bog's Mill (1900) and Redhall (1902) were taken to increase output. At Redhall a turbine with an 18 foot head was installed in a new brick building upstream of the old mill but using the same lade and this doubled production to about 10-20 tons per week. The sawdust was shipped from Grangemouth by barge to Stoneyport on the Union Canal, the former quarry loading point just west of Slateford, and the wood flour was sent out from Kate's Mill siding on the Balerno Branch railway en route to Kirkcaldy. A powerful mill at Blantyre on the Clyde was also used (1906) and later (1926-28) Arrat's Mill on the South Esk in Angus was occupied to serve the Dundee trade. A further mill at Carmyle was also involved until 1936.

Gideon Walker died in 1912, but Allan was eventually joined by his sons Allan and Lawrence, who were born at Bog's Mill in 1926 and 1928. Grinding sawdust was a risky business and there were many fires. Bog's Mill burned down in 1924 and a blaze at Redhall in 1958 caused production to move back into the old mill there. The biggest calamity, however, came when Bell's Mills blew up in 1972. Ironically, wood flour had been added to T.N.T. as a stabilizer during the war years. Lawrence Walker and Bob Johnson were severly burned in the explosion and Lawrence regained consciousness lying in the ruins of the mill, which never worked again. Lawrence was in hospital for a month at Bangour until his skin began to recover. He returned to work at Redhall until 1980 but still bears the scars of masonry chips from the explosion. Lawrence is understandably proud to be the last of the Water of Leith millers and the guardian of much local milling lore.

Bob Johnson's brother-in-law Archie Cairns also worked at Bell's Mills in their last days and recalls them with affection. He remembers the back-breaking task of clearing the dam-head in summer, building a coffer dam to divert the water while the tree trunks, silt and other debris were removed from the weir and the lade. At that time the mill generated electric power from the lade and had its own wood-working shop with circular saw, lathe and drilling table. "Strange as it may seem, I can still recall the sound of the mill even after all these years: the water spilling from the buckets, the gears creaking and groaning and the millstones rumbling as they turned!"

Table 1. Some serious mill fires and disasters

1655 Murrayfield (Dalry) Paper Mill
1682 Canonmills Paper Mill
1800 East Mill Bank Barley Mill, Currie (approximate date)
1836 Stenhouse Mill
1867 Bog's Mill
1885 Slateford Mill
1890 Kate's Paper Mill
1901 Todd's Mill (Stockbridge) exploded
1909 Balerno Bank Paper Mill
1916 Kirkland (Richardson's) Board Mill
1920 Newmills Barley Mill
1924 Bog's Mill
1954 Spink's Bung Mill, Malleny
1956 Jericho (granary) at Dean Village
1958 Redhall Mill (woodflour plant)
1972 Bell's Mills exploded.

Watermills of the Bavelaw Burn

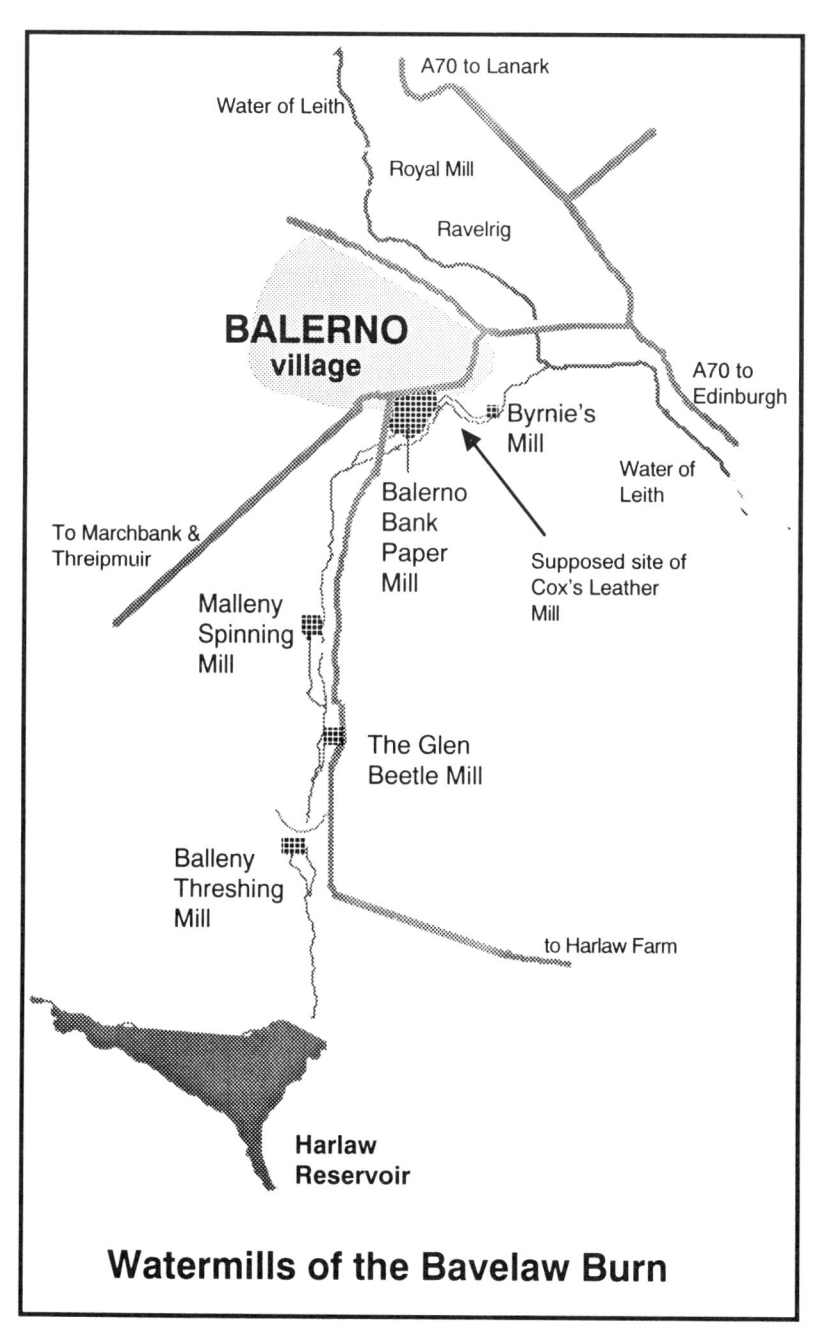

Balleny Mill on the Bavelaw Burn as it is now, following transformation from ruin to family house in 1993. The wheel has been reduced to its famework, the upper half of the wheel house and the trough bringing the water to the wheel have been removed but the lever which stopped or started the flow of water onto the wheel can be seen poking out of the wall higher up. This was operated from inside the mill. The small archway for the tail-lade can be seen at the foot of the wall beside the (new) drainpipe.

CHAPTER 5.
A check list of mills from source to sea

Six figure numbers are National Grid references on square NT, the river being covered on the Ordnance Survey's Landranger Series Sheets 65 and 66. The first three figures of the reference are eastings (west-to-east position) the second three northings (south-to-north position) and the margin of each map explains how they work. They are more useful and accurate in the country areas of Sheet 65 than in built-up Edinburgh but, in combination with the site description (e.g. 100 yards upstream of Gorgie Road), should help orientate any reader who wants to check the sites out. At some of them, of course, all traces of the mill have disappeared.

Leithhead Mill NT114637

As the name indicates, this mill is at the top of the river, some 5 miles upstream of Balerno and close to the Lanark Road, in Kirknewton parish and perhaps serving that village. lt had a marriage stone dated 1725. Paper-making (1 vat; George and Andrew Gourlay) and corn milling were powered by the same 18 foot wheel in 1825-1832 but after that only corn and barley were milled. A second wheel on a subsidiary watercourse across the lane may have been used for threshing. The weir and lade are traceable, the lade giving sufficient head for an overshot wheel. The mill ceased to work in 1924 and was converted into a house, much altered in the 1980s, when the date stone disappeared.

Cockburn Mill NT143657

A small 19th century estate sawmill, powered by an undershot wheel on the Cock Burn in the hamlet of Glenbrook, WSW of Balerno. The weir was destroyed by a flood in the 1960s and the mill has not worked since.

Royal (Ravelrig) Mill NT157644

Opposite Larch Grove but on the north bank of the main river, and possibly part of the Ravelrig estate at one time, this grain mill was operating in the 18th century at the time of a damaging flood (1739) but was short-lived. Contrary to one local account

Here today and.......Balerno Bank Paper Mill in 1968 at the height of its success. The only surviving parts are the mill office on Mansfield Road, the small building with the pointed roof in the right foreground, and the house beside it. (from the company history "Galloways of Balerno")

there is no evidence that it made paper. One miller, Alex Rankin, was buried in Currie kirkyard in 1750. The circular kiln survives in ruined form as the only trace of the mill. It could almost be a broch at first sight and may have been adapted later as a lime kiln.

Bavelaw Mill Farm NT151625

Now only a farm name, this may have been a seasonal mill, perhaps for threshing. The Bavelaw Burn here (2 miles SSW of Balerno) is very small. The mill was shown on John Lawrie's map of 1766. There were plans to change the course of the burn in 1875 but it is not clear whether the mill was working then.

Balleny Threshing Mill NT175654

Situated just below Harlaw Reservoir on the Bavelaw Burn, this was a simple single-storey farm mill used for threshing and originating in the mid-19th century. It had a short but interesting lade which turned through 90 degrees to reach the 18 foot pitchback wheel via a length of concrete pipe and then a wooden trough or launder. Almost derelict in 1993, it was then converted into an attractive dwelling by a local family, retaining the metal framework of the wheel in place (page 42).

Malleny Mills NT172656 and NT169657

Two mills were set up here on the Bavelaw Burn by the Edinburgh & Leith Ropery Company and gave rise to a small community separate from Balerno and lying just to the south. The spinning mill (1805) processed flax into thread and a beetling mill (1825), with a 25 foot overshot wheel, was added just upstream to finish the sail cloth. Shallow retting ponds for the flax, where the soft parts of the plant were allowed to rot away, are still visible slightly to the west; the rectangular stone-edged one was later used for water storage by Balerno Bank. There was also a gas plant for bleaching either the yarn or the cloth, probably with chlorine, and a green where the product was dried. The beetling mill (currently a furniture workshop, The Glen, on Harlaw Road) became a school for Malleny children in 1841, then a grain mill. The spinning mill (Hay's Mill in 1850) became Spink's Bung Mill in 1883, and produced wooden bungs for beer barrels, plus chair and table legs. An old postcard shows a fine three-storey building with piles of timber stacked alongside. A wood south of Currie is referred to as Spink's Wood by some local residents. The mill burned down in 1954, was demolished, and the site is now covered by housing.

Balerno Bank Paper Mill NT 164663

Expansion of this mill (built 1805 by Robert Walker) and realignment of Harlaw Road saw the end of the smaller **Bayne's Townhead (paper) Mill** by 1830. Owned by Hill, Craig & Co in 1840-60, Balerno Bank was served by the Balerno branch line although the terminus at the goods yard was a little way to the north, across the river. There were two 18 foot wheels, probably overshot. The mill burned to the ground in 1909 and was bought by John Galloway (ex-Portobello Paper Mill) in 1924. Under Galloway's inspired but demanding direction it had a successful steam and electric-powered career with regular investment in new processing (about £2 million from 1945 to 1968) and a capacity of 12,000 tons per year by 1968 when there were 400 employees (see page 44). The one-time heart of Balerno was demolished in 1989, after serving as a paper-coating plant in its last few years and the site was redeveloped for housing in 1996. The attractive weir and sluice survive on the Bavelaw Burn and beside the driveway to the Kestrel Hotel. The red sandstone mill offices on Mansefield Road, dated 1924 with the initials JG Co., have recently been converted into a dwelling.

Byrnie Mill NT165664

Established as Helen Logan's mill in 1799, this was originally a small (one vat) paper mill on the left bank of the Bavelaw Burn not far from its confluence with the main river. It was not listed as a paper mill in 1852 and may have become a sawmill at that time. Later it became a water-powered furniture workshop, Smith's Sawmill and Joinery, the wheel working until after the Second World War. Bill More, born in 1922 and still living in Currie, recalls that as the junior of the two apprentices in 1937 his first task early each morning was to clear the grid at the weir of leaves and other debris so that the water could be turned on. The firm always managed to work in summer despite the low water levels. The building comprised a sawmill downstairs, driven directly from the waterwheel and a joinery upstairs with belts driving the drills, lathes and smaller saws there. The work was mainly large-scale joinery for housing, with a contract for packing crates for an Edinburgh firm. Bill left to join the Royal Marines in 1939 and returned only briefly after the war. The workshop was in use in 1970 under electric power, but was then demolished and the 12 foot iron pitchback wheel standing beside the burn was broken up despite being in saveable condition. The weir was upstream, alongside Balerno Bank and the water was piped downstream to the wheel.

Byrnie was just across the Bavelaw Burn from Malleny House (1478), but there is no evidence to support the suggestion that it might have been the estate mill prior to its days as a papermill. A curious use of waterpower at Malleny was mentioned in one of

James Thomson's poems, "Braes of Malleny":
> *The wimplin' burnie murmurs through*
> *The underwood, whiles hid from view:*
> *Now rushing in abrupt cascade,*
> *Refreshes the surrounding glade.*
> *Then disappears, but is not lost,*
> *It turns the spit that bears the roast.....*

but no trace of this ingenious arrangement survives today.

One other water-powered enterprise seems to have operated for a short time just upstream of Byrnie. Soon after John Cox bought Gorgie Mills in 1805, his son Robert set up a mill on the Bavelaw Burn for "the waulking of leather." John Tweedie put the site above Byrnie but below the present Harlaw Road bridge and thus the intake for Balerno Bank. It must have needed clever management of the water supply and the mill does not seem to have lasted long once the Cox family were established at Gorgie.

Newmills NT167670

John Skene of Curriehill Castle owned this grain mill in 1604. By 1850 it had two 16 foot wheels, possibly overshot, each driving two pairs of stones for barley. It was producing oat meal under the ownership of William Ross in 1920, when it burned down. The mill house survives beside Lanark Road West; the mill occupied the garden area. There is no weir now but the head and tail lades can be traced, the latter lined with careful stonework where it passes under the railway embankment.

Balernoch Waulk Mill NT173672

This is the oldest mill site above Colinton (1376), on an elbow of the river. It was later adapted for other uses (a distillery in 1845, recently a piggery, now a private house) while the lade was extended to serve the next mill down. It is commemorated in the name Waulkmill Loan for the short lane down from Lanark Road West.

Kinauld Tannery (Balerno Paper Mill) NT176674

Sited on the north bank, just in Currie, but set up in 1788 as Balerno Paper Mill for Nisbet & McNiven, this was a large concern with 6 vats. There were two 18 foot wheels in 1850, one

The kiln at Currie viewed from the south

developing 36 horse power. Paper-making ended in 1882. Later it was Durham's glue works and then a tannery (J Hewit & Son from 1913) with its own short railway siding. A turbine was in

East Mill Snuff Mill from upstream with the weir for East Mill Bank Mill in the foreground and the ralway bridge showing behind on the left. How could the powers that be allow such a building to be demolished? (Photo: Royal Commission on the Ancient and Historical Monuments of Scotland.)

use until about 1950, making it one of the last water-powered sites on the river. Hewits still occupy the premises, producing specialist leathers for book-binding. The weir was sited above Balernoch on a sharp bend in river, the lade passing under the railway line. A deep tail-lade is traceable almost down to Currie Bowling Club.

Currie Mill NT182678

This grain mill (1506) was on the north bank beside Currie Brig. In 1785, George Davidson, who lived at Wester Currie Farmhouse (now the Riccarton Arms), had agreements with the Craigs of Riccarton and with the owners of Hermiston to grind their corn and to receive ale brewed at Hermiston. It was

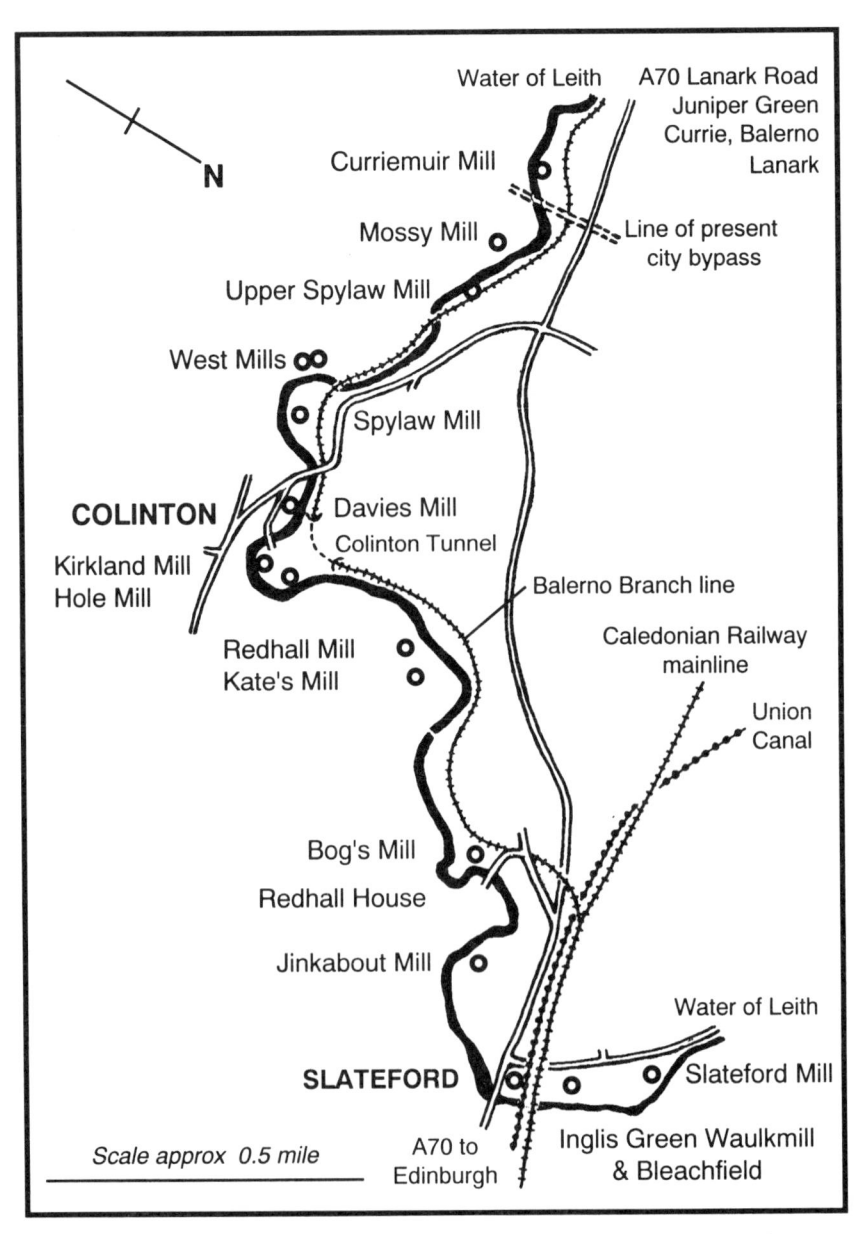

*The mill sites around Colinton: the Balerno Branch railway
line is now the Water of Leith Walkway*

probably the same ale that was sold to bypassers from the house beside the brig by Marion Cunningham "for forty years at least" according to another James Thomson poem. The mill was no longer operating in 1850. The kiln remains as an attractive roofless ruin beside the bridge (page 48). It has changed little since the Westwood drawing of 1808 but now has only traces of stonework beside it as relics of the mill and cottages. The site of the weir is just discernible upstream.

Kinleith Mill NT189679

Kinleith was a waulk mill (1618) on the south bank, converted to paper-making in 1792. In 1850 its 10 x 23 feet wheel developed 36 horse power and in 1878 a chimney 280 feet high was added to carry away the smoke from its steam engines after complaints from the parochial board. By 1882 Kinleith was the 5th largest paper mill in Scotland with nearly 400 employees, mainly local residents of Currie and Juniper Green, producing 287 tons of paper each month. In 1896 the owners of Kinleith, the Bruce family, built their mansion, Braeburn, later the Glenburn Hotel and now adapted as housing, on the hillside above the hamlet of Blinkbonny, where some workers and the mill manager lived on the north side of the road.

Kinleith was a major user of the railway, with its own electric shunting engine working a complicated arrangement of sidings within the mill site. In 1957 some 1,252 wagons were dispatched via the siding, nearly all containing paper. The mill closed in 1966 "effectively sealing the fate of the Balerno Branch line" (Donald Shaw), with piecemeal demolition continuing in 2000. The high weir and first section of the lade remain in good condition.

East Mills NT192682

The grain mill (1625) lay on the south bank downstream of Mutter's Bridge carrying the road to Blinkbonny over the river (and later the railway). The weir is just upstream of the bridge and the lade passes through its own culvert and then continues close to the Walkway. A wheel pit (18 foot) about 70 yards beyond the bridge shows the grain mill site. It worked until 1900. The mill house survives, now much extended. In 1749 the first snuff

Upper Spylaw Mill from the south and slightly upstream. The water wheel was in the basement behind the single lowest window in the gable. The downstream side of the mill has a courtyard with a stair to the upper storeys.

mill on the river was erected farther down the same lade (just above where the railway bridge crosses the river) by Abraham Ferrier and James Thomson. Photographs (see page 49) and an 1899 drawing show a pretty stone building with crow-stepped gable ends, a wooden extension overhanging the river and its own footbridge. It had two 14 foot wheels in 1850 but only one wheel pit can be traced in the lade. The mill worked until 1920 and was demolished after brief use by Edinburgh Scouts.

East Mill Bank Mill NT193683

Situated opposite the snuff mill, on the north bank of the river, this red sandstone barley mill (James Watt,1800) had a short life, being disused in 1850, and was apparently destroyed by fire (1820?). The ruin can still be seen. It must have suffered from lack of water due to the two mills opposite taking much of the flow into their lade, but perhaps it was under the same ownership.

Woodhall BoardMill NT196683

On the north bank, this was a waulk mill in 1747 but was converted to paper in 1792. In 1850 it had one wheel of 18 x 5 feet rated at 20 h. p. In 1872 damages (£50) were paid to the owner when construction of the railway stopped the mill working for two weeks. It became a major user of the railway and later had sidings and storage at Juniper Green Station about 200 yards downstream. Latterly the mill processed waste paper to make board for packaging as part of the Inverleith Paper Group. It closed in 1984 and was demolished, the long narrow site still laying waste in 2000. The weir and elaborate sluice (page 14) survive.

Woodhall Bank Mills NT199687

Another pair of mills (both 1763) shared the same lade on the north bank, with the weir just downstream of the Board Mill and about 100 yards upstream of Juniper Green Station at the foot of the Brae (behind the current Post Office). The lade passed under the Brae, behind the station and drove first Wright's Grain Mill then Watt's Snuff Mill, the latter nestling beside the railway track, its top floor barely level with the trains. Nothing remains of the grain mill and the foundations of the snuff mill are rapidly disappearing under the leaf-mould but two tail lades can be traced at the riverside, so there must have been a bypass sluice to permit independent working. The snuff mill worked until 1940. A draft 22-year lease between the laird, Sir William Liston Foulis, and the snuff miller, Robert M. Watt, in 1897 shows a rent of £16 with obligations to maintain the damhead (weir), lade,

Spylaw House: James Gillespie's mansion fronting his snuff mill

water wheel and spur wheel and to insure the property with a reputable firm (against fire damage) for at least £150. The previous tenant had been John Watt, Robert's father, described as "recently deceased". The miller in the barley mill is named as Archibald Walker and a kiln is mentioned, presumably part of the barley mill.

Curriemuir Mill (Inglis Grain Mill) NT202687

Opened in 1704, this mill just upstream of the City Bypass still dries grain at present, although nearly all traces of water power have gone - sharp eyes will detect the weir site at the bend of the river upstream. The mill produced oatmeal for many years. In 1850 it had two wheels, both 4 x 14 feet and 16 h.p., with three pairs of stones for each. Several applications to redevelop the site for housing have been refused over the last 10 years, and another is pending as this is written.

James Gillespie's Spylaw snuffmill from upstream. Compare this simple building with its grand frontage (opposite). The lade must have entered the mill about where the single lower window is now placed.

Mossy Mill NT206688

A waulk mill (1664) on the right bank, originally run by the Mosey brothers, hence the name, the mill was making paper by 1838. The 1850 wheel survey shows three wheels of 18, 14 and 12 feet, the largest rated at 26 h.p. The watercourse was unusual: water left the lade at 90 degrees to drive the wheels within the narrow building and then fell almost directly into the river with no tail lade. The impressive weir and the beginnings of the head lade survive, together with a fine manager's house on the top of the bank and the route followed by workers, down steps from Lanark Road, by tunnel under the railway and via a lighted footbridge over the river, can still be seen. After the paper mill closed (1972) the site was used by a waste disposal firm until demolition for housing late in 1999. Among the foundations, after demolition, it was still possible to see cavities for wheels two and 6 feet wide and tanks up to 15 feet in diameter.

Upper Spylaw Mill NT207688

Reached by a lane off Gillespie Road in Colinton, this elegant and interesting four-storey building on the north bank dates from 1682, the fourth oldest papermill in Scotland, under James Lithgow. A snuff mill in 1765, its owner William Reid was a rival of snuff miller James Gillespie at neighbouring Spylaw. The mill's reputation for also serving as an inn favoured by smugglers was used as a pretext for a search by excise men tipped off by Gillespie and accompanied by his workers so that they could inspect the new features of the milling machinery. There is an interesting outside stair up from the courtyard at the rear, which might have served the drinkers and in fact the smuggling charge was supported when large quantities of tea and brandy were found hidden in the upstairs rooms in 1776 *(Edinburgh Advertiser)*. Upper Spylaw had one 16 foot wheel in 1850 but the mill had become a dairy, with its own railway platform, by 1880 and is now a private house. Both head and tail-lades were very short and the weir was very close to the Mossy Mill outflow.

West & King's Mills, Colinton NT211688

These mills take us back onto the south bank. This excellent site with a good head of water had 3 mills in 1688, for waulking, flax and grain. By 1791 only grain and paper mills were operating, with the paper-making supporting 300 people. Members of the King family (hence King's Mills) milled grain for a hundred years and are buried at Colinton Kirk. In 1850 Westmills had one 12 h.p. wheel (12 x 2 feet) and King's had two, both 17 x 3 feet (8 h.p.). By 1909 papermaking had ended and A & R Scott (of Scott's Porage Oats) took over until the firm moved to Cupar in 1971. The building on the north bank (used by the travel firm Globespan in 2000) was also part of the complex, the mills' railway siding passing through it so that wagons, but not locomotives, could be shunted across the river on a separate bridge, which still survives. The main site was used for dairying, and later antique sales, but demolition followed. The weir was declared unsafe and removed in the 1980s. The lade is traceable and some mill workers' houses survive. The range of more than 20 millstones, mostly edge-runners, arranged on two levels beside the Walkway, came not from a local mill but from an Edinburgh coffee factory at Abbeyhill in the 1970s.

Spylaw Mill NT213689

Famous as the mill of James Gillespie, a renowned Edinburgh snuff merchant and benefactor, the 4-storey mill (best seen from the riverside, see page 57) dates from 1650. Once installed there in 1759, Gillespie was soon wealthy enough to build an impressive mansion across the front of his mill (1779; see page 55) and the sight of his coach inspired the couplet:

"Wha would hae thocht it
Noses had bocht it?"

After Gillespie's death (and burial at Colinton Kirk) milling was continued by the Richardson family but they moved their operation to Kirkland Mill beside Colinton Kirk and Spylaw was out of use by 1850. The building is now converted into flats in what is Spylaw Park, overlooked appropriately by Gillespie Road, and by a new bridge (1873) over the river and the railway. Gillespie's money founded a school that still exists and bears his name, a hospital in Bruntsfield, and a row of workers' houses in Spylaw Street, Colinton.

Davie's Mill NT215691

This was another ancient grain mill (1509), on the south bank, with access and mill houses beside the original Colinton Bridge at the foot of Spylaw Street, but there is confusion over the site of a magnesia factory in Colinton: was it here for a time (on the same lade as Davie's according to John Tweedie) or on the burn at Bonaly a mile to the south? In 1850 two 16 foot wheels were used for corn and barley. Some buildings currently survive, much altered, and a line of stones may mark the weir, with the outflow just above the bridge.

Kirkland Mill NT217692

This mill stood on the north bank beside the kirk, on the site of a mansion owned by the Borthwick family. In 1777 it was a lint (flax) mill, then a meal mill (1829), then Richardson's Snuff Mill (1846; also called Spylaw) with one wheel 15 x 7 feet (16 h.p), and finally it made board for book-binding. lt was then a large stone and timber building with a tall chimney but burned down in 1916 and the site was used for a new graveyard. The line of the weir can be seen just below the bridge and the small house that may have served the mill manager was later used by a grave digger and had been a school.

Hole or Hailes Mill NT217693

This was the original Colinton waulk mill, mentioned in 1226 as the property of Thomas de Lastalrig, but later (1625-1880) a grain mill. lt lay behind the manse and the young Robert Louis Stevenson visited it from the manse garden when his grandfather Dr Lewis Balfour was the minister (1823-1861), recalling his fascination later in the poem Keepsake Mill, published in *A Child's Garden of Verses*. The idea expressed there that the wheel will continue working for ever was rudely denied by demolition of the mill even before publication of the poem. A fine painting now in Colinton library, by the artist John McWhirter (of Mossy Mill and Inglis Green Bleachfield), shows the wheel house on the river side, sluicegate for a short lade and 3-storey mill. The wheel was 16 x 4 feet in 1850.

Redhall Mill NT214697

One of several mills on the Redhall estate, on the south bank half a mile below Colinton Bridge, this was a paper mill in 1718 and later (making bank notes in 1769) but a barley mill in 1742. lt was reconstructed in 1803 (recorded on a stone on the west side) and in 1850 had two 16 foot wheels (8 h.p.). After 1902 it was used by the Walker family (of Bell's Mills and the Dean) and worked up to 1980 using a turbine with an 18 foot head to grind wood flour for linoleum manufacture using two pairs of stones. This enterprise occupied a new red brick building but after a fire in 1958 moved back into the old mill. The V-shaped apron weir remains impressive and the lade still carries water towards the mill. It reappears on the east side and drives a new undershot wheel on the wall of a two-storey workshop. The wheel is a novelty and drives no machinery but is the first to turn on the river for almost 30 years (page 80). The square-sided kiln shown in James Stuart's drawing of 1900 has disappeared and the mill itself was converted into two houses in the 1980s, with one retaining the pit for the breast wheel in the middle of the living room. This remains an attractive group of buildings with its cottages. The tail lade went on to Kate's Mill but now turns sharply back to the river beside the garden of Kate's Mill Cottage.

Kate's Mill NT214698

Developed on a waulk mill site, this mill was converted to paper-making in 1787 and was reputedly named after Kate Cant, wife of John Balfour, formerly of Bog's. It prospered and shipped its paper first via Stoneyport on the new Union Canal in 1822 and then (under David Chalmers' ownership) via the railway. It had one 26 h.p. wheel of 12 x 14 feet in 1850 plus steam power. By 1865 it was the 20th largest paper mill in Scotland and 71 people lived there, but it burned down in 1890 and was never rebuilt. Some stonework is visible behind Kate's Mill Cottage, with more below ground level, and there are hints of a tail lade beside the river.

Bog's (Boag's) Mill NT217702

On the north bank, half a mile upstream of Slateford, this was a waulk mill (Vernour's) in 1598, converted to corn in 1631 and was leased by Nicol Lithgow for paper in 1717. In 1735 it made paper for Bank of Scotland 20 shilling notes, the bank officials staying in Millbank House, an elegant house which survives just to the west (page 25). Gavin Hamilton, one of three partners in 1756, had negotiated Edinburgh's safety with Charles Edward

Stuart at Graysmill in 1745. Later, the sole tenant John Balfour, refused permission to extend by his landlord George Inglis of Redhall, leased the waulk mill upstream and converted it to paper-making as Kate's Mill. In 1850 Bog's had two 20 foot wheels, one undershot, one breast-shot and milled barley and spices (including pepper, curry, ginger, cinnamon, mace and cassia, according to Geddie) for several generations. An attractive line drawing by Henry Westwood, dated 1896, reproduced in John Tweedie's *A Water of Leith Walk* shows the kiln, lade and undershot wheel standing outside the mill and a photograph (1890) also shows white-walled cottages closeby. The Walker family occupied Bog's in 1902 when they needed to increase their output of wood flour and the two pairs of stones were put to this use. The mill burned down in 1924. Its ruins and the row of cottages were demolished in 1970. Lades can be traced with difficulty and the mill site is occupied by a private house.

Jinkabout Mill NT219704

This was Lumsdaine's corn mill in 1506. John Reid, owner of the Edinburgh Gazette, made paper here on the north bank of the river for 25 years from 1714. It was advertised for sale in 1755 but demolished soon after in the making of the walled garden for Redhall House when George lnglis bought the estate. There is no trace today.

Slateford Mills NT220708

This complex on the north bank (more west here) straddled the Lanark Road, with the weir just upstream of the road bridge (1764). The site was then bisected first by the canal (1822) and then the Caledonian Railway (1847), so that the river is crossed by road bridge, the splendid 8-arched canal aqueduct and railway viaduct within a hundred yards. A waulk mill existed in 1659 and may be the two-storey building shown on engravings as standing on the lade just south of the aqueduct, on the 1893 O.S. map and on a photograph in 1927. Inglis Green Bleachfield started in 1773 and developed into a major concern for 200 years, dyeing and bleaching linens and cotton, weaving and dyeing woollens and later operating as Inglis Green Laundry (A & J McNab). The

Inglis in the name comes from the laird Sir George Inglis of Redhall. The site included Graysmill, a farm and grain mill, where Charles Edward Stuart stayed briefly in 1745. This mill was eventually replaced by Slateford Mill for paper by 1893 at the lower end of the lade. The 1850 survey lists two wheels 14 x 4 feet and 16 x 6 feet 'for washing and waulking', with another 18 x 5 feet at Graysmill for flour and barley but no paper mill at that time. The western half of the weir has been covered up and the culvert for the lade to pass under the Lanark Road disappeared during road widening (1967). The laundry site was also redeveloped in the 1960s and only traces of the older buildings remain. The Village Inn, down lnglis Green Road, is on the site of the paper mill and may incorporate fragments of the mill.

Stenhouse (Saughton) Mills NT218714

The river turns a right angle at Longstone, beside the Inn and just beyond here are the remains of the weir for the next mill on the north (west here) bank, an ancient foundation gifted to Holyrood by King David I and still in the possession of the abbey in 1511. The mills were the nucleus of a small village (population 95 in 1841) which was later part of the lands of Saughton and the mansion house at Saughtonhall (now Saughtonhall Park). Grain milling went on until around 1900 but was accompanied by other enterprises at different times, including waulking, brewing, weaving, snuff-making (in 1740, William Reid, later at Upper Spylaw?), pasteboard and cardboard manufacture, hosiery, lace and threshing. There were at least two mills (easter and wester) most of the time and in 1850 three wheels, two for corn and barley of 20 x 5 feet, and one for threshing 16 x 4 feet.

In *Historic Corstorphine* A S Cowper relates that several families occupied the mills for many generations, including the Stenhopes (1511-1605), Brouns (1661-1717) and Cleghorns (1751-1800). The 16th century mansion house (restored 1937-39) was enlarged in 1623 by Patrick Eleis, an Edinburgh merchant, Town Councillor and Royal Commissioner who recruited weavers from Flanders. A datestone over the door on the eastern side has the initials PE with a fine coat of arms and the inscription "Blisit Be

Stenhouse Mansion from the north.

God for Al His Giftis". Hand loom weaving continued here for some time and an inventory of William Murray's possessions in 1739 lists '6 Holland looms, 4 Dutch looms with Holland webs in them, 4 Scots looms with a web of Dornick in one, and 2 pair reels for winding yarn.' This splendid building, now sadly hemmed in by unsightly modern developments, is home to the conservation service of Historic Scotland. In 1700 there was an extensive garden to the north, with a total of 142 apple, pear, cherry and plum trees plus nut bushes, gooseberries, raspberries and currants. There was also a doocot which lasted at least to 1845. Bits of old mill building survive, with cottages up on Stenhouse Mill Wynd, but part of the site was occupied by a greyhound stadium in 1937 and now has mixed industrial use. The new footbridge for the Water of Leith Walkway is on the site of the mill bridge.

Dalziel's Mill NT218717

Latterly, the tail-lade from Stenhouse Mills went under Gorgie Road and continued on the west bank almost to the footbridge at the end of Ford's Road but in the period 1576-1796 Dalziel's grain mill occupied the north bank close to what is now Gorgie Road. lt was part of the lands of Saughtonhall. A.S. Cowper's *Historic Corstorphine* recounts a complicated and often violent family dispute around 1600 that had Archibald Dalyell imprisoned in Edinburgh Castle for attacking his own tenant millers and their families "with ane drawn sword" and on a later occasion scourging James Brigtoun for half an hour with "sword belts and hingers" so that "none of his body was frie of bloudy and bauch straikis." On release Archibald returned to the mill to destroy its wheels, axletrees and millstones. After further bloody violence Archibald's father Robert was also imprisoned. In 1603, Archibald escaped further action by the Privy Council when he claimed to have caught the rebel Neil MacGregor of the proscribed MacGregor clan. There were disagreements with Stenhouse Mill over flooding in 1576, and again in 1753, and the last records of the mill are about 1799, with the tenancy for the Trelss family.

Gorgie Mills NT226722

Of ancient, possibly even 13th century, foundation, Gorgie became a mills complex on the south bank of the river, straddling Gorgie Road with a grain mill north of the road and (from 1789) an extensive glue works on the south side. Cox's Glue Works lasted almost 200 years and by 1850 Gorgie had six wheels: two of 9 x 6 feet were for corn and barley; two of 10 x 3 feet were for grinding bark (for tanning) and washing hair and two of 10 x 10 feet were "float wheels" for pumping water. The corn mill building survives, occupied today by Didcock's Furniture Company. The glue works site was cleared in 1969 and is now occupied by BT House (1973).

Cox's was soon a major local employer, with 100 workers by 1856, when Cox petitioned parliament. He complained that discolouration of the Water of Leith by peat was darkening the colour of his glue and the works was granted its own spring water supply (72,000 gallons per day) piped from The Edinburgh Water Company's Torduff Works. Later, Gorgie had its own railway siding. John Cox was a great benefactor, erecting the Edinburgh Royal Patent Gymnasium at Canonmills, boating, skating and curling ponds at Craiglockhart and financing the dome on the Old College of Edinburgh University. The weir and entry to the lade survive alongside Gorgie Road near the junction with Balgreen Road and there are still traces of the lade beyond the mill where it went under Stevenson Road and approached the river along a neglected narrow strip of land. Before 1900 the lade continued on to the Dalry Mills, passing close to Damhead Farm, a name suggesting a milling use at some time.

Dalry (Murrayfield) Mills NT227727

Served by the lade from Gorgie, these mills operated from 1576 (with Mungo Russell) to the 20th century, and in 1595 housed Scotland's first paper mill. Papermaking and grain milling went on together but in 1771 the engineer John Smeaton was called in to advise on water wheels for a complex of three mills making oatmeal, snuff and flour. Smeaton proposed an 11 x 4 feet overshot wheel but by 1854 there were only two wheels, both 16 x 6 feet, for flour, corn and barley. On the evidence of O.S maps, use of water power seems to have ended around 1900 when the lade was cut off. Mungo Russell's Roseburn House, dated 1582, survives, with a later (Victorian?) mill house and some cottages on Roseburn Street. The mills themselves (John Hunter's Murrayfield Corn Mill in 1908, when Murrayfield was a polo ground) occupied the area close to the Murrayfield Rugby Ground tumstiles on Roseburn Street. Bypass and main tail lades rejoined the river in what is now Roseburn Park.

Coltbridge Mills NT229733

Coltbridge was a mill village until the mid-19th century, the mill(s) on the river's north bank lying just downstream of the old bridge on what is now the Glasgow Road. Murrayfield, including Coltbridge, was owned by the Nisbet family of Dean. Earlier tenants (1729-1790) were skinners and in 1810 a saddler, Robert Paterson, owned 3 mills, so more than just grain milling must have gone on, but by 1850 there was one 10 x 4 feet wheel used for flour. The weir (removed about 1920 after it was blamed for flooding) was just below the road bridge with a short lade; the tail

Part of a view of Dean Bridge (1880?), showing Mar's Mill, with its wheelhouse jutting out across the Great Lade. The lade went round the bridge pier and served Greenland Mill, out of sight to the left, before travelling on to Stockbridge, Canonmills and Logie Green. The massive 6-storey building beside the mill is Jericho, the city's girnal of 1619. Notice the low flow in the river itself - most of it must be flowing down the lade.
(George Washington Wilson Collection, Aberdeen University Library.)

lade can still be seen emerging beneath the bridge for the Granton branch of the Caledonian Railway. In 1894 J & W Arbuckle were the millers but they went bankrupt and Allan Walker took over. The mill was restored for housing in 1983. Coltbridge House survives.

Bell's Mills NT237736

One of the river's last mills to be worked by water, on the north bank by Belford Bridge, was demolished after an explosion in 1972 when it was making wood flour for the linoleum industry (see Chapter 3, page 38) using an 18 foot breast wheel and a turbine. At that time it also generated electricity for its own use. Several generations of Walkers milled here, including Gideon from about 1890, his son Allan and then grandson Lawrence, who was injured in the explosion. They lived in the mill house beside the mill's steep cobbled lane. The site was very old, possibly 12th century, and the Nisbets of Dean (1605-1845) were owners. John Cox used a small waulk mill here in 1798 before going to Gorgie and the Mills (always plural) may have had several uses at one time. Indeed there was a population of 67 people in 1890. The 1850 survey shows two 16 foot wheels.

Following the explosion, the site was occupied by a car saleroom and workshops and then redeveloped for the Dragonara (now Edinburgh Hilton) Hotel. The elegant former granary was retained as part of the hotel. The long oblique weir is reasonably intact and water runs down the lade, most of it re-entering the river by a relief sluice, but the final portion of the lade provides a water feature in front of the granary. A small ruin upstream by the river is a former mill, disused since about 1807 when the height of the lade was raised to increase the power available.

The Dean Mills NT240740

Grain mills clustered here on both sides of the river were the bread basket of Edinburgh and were run by the Incorporation of Baxters (Bakers), the oldest and strongest of the city's trade guilds. The mills existing in 1128 were given to the Abbot of Holyrood by King David I and milling seems to have developed steadily from then onwards, the importance of the settlements

The West Mills at Dean Village. Note the ports at water level on the nearest mill and on the right of the small lawn to return water to the river.

accentuated by their position beside the vital river crossing on the route to Queensferry. The area now called Dean Village was once called the Village of the Water of Leith, and Cumberland Hill, writing in 1874, distinguished Dean Village as a separate hamlet more distant from the river on the north bank. By the 17th century the Baxters had 11 mills, with their own girnal (granary) alongside the mills and even at the beginning of the 19th century their emblem, a sheaf of wheat, was placed on mill buildings (Bell's Mills' granary and West Mills).

There are two high weirs: the upstream one powered mills on the north side of the river and the lower one powered mills on the south side. Most of those on the north bank, said to total 8, plus Hamilton's Brewery and Hufton's Chemical Works, were swept away to make Legate's Skinworks in 1843, leaving just the lowest pair, the West Mills (1805). These handsome buildings, (seen above and on the cover), survive as flats following conversion by

Link Housing in 1973. In 1850 the West Mills had a powerful wheel each, 14 x 12 and 18 x 9 feet worked by a 14 foot fall, and the paired ports by which the water returned to the river are visible on the river side of the southern mill and on the downstream side of the northern mill (where one has been bricked up).

Immediately below the West Mills is the lower weir, some 12 feet high. This fed three mills in succession, Lindsay's (1556) being virtually at the end of the weir, then Mar's, then Greenland beyond the arches of Telford's soaring Dean Bridge (1831). ln 1850 these mills had one wheel each, Greenland's the largest at 20 x 4 feet (12 h.p.). Beyond Greenland Mill "the Great Lade"continued east, running along a wooden trough supported on stilts. lt crossed Kerr Street to reach Stockbridge Mill, continued to Silvermills, then to Canonmills and finally Logie Green Mills before returning to the river, a distance of over 1.5 miles. A photograph of Dean Bridge by George Washington Wilson (see page 65), taken from the north side of the valley, shows Mar's Mill, with its wheelhouse jutting out and the lade passing on both sides of the bridge pillar, and closeby the Jericho girnal (1619). Greenland Mill was occupied by Gideon Walker in 1880 in the early stages of the Walker family's wood flour making.

The lade was removed around 1890 after earlier complaints by Dr Henry Littlejohn, Edinburgh's first Medical Officer of Health, about its insanitary state. The only trace of it now is the blocked up arch beneath the south end of St Bernard's Bridge. The lower part of Lindsay's Mill, best seen from the riverside below the weir, survived demolition in 1937 and on what must have been the second of its original five storeys three French mill-stones are arranged, leaning together to display their grinding surfaces. A proposal in 1996 to renovate Lindsay's Mill failed to win the support of local residents.

Stockbridge Mill NT246746

Farther downstream on the south bank and taking next use of the lade after Greenland, this was another grain mill, situated about 100 yards from the river east of Kerr Street. lt is poorly recorded but may have belonged to the Nisbet family. The miller

The Canon Mill from Eyre Place

in 1760, Adam Smith, employed a trace-horse to haul carts up the steep slope from the ford (Cumberland Hill). A steam engine seems to have been introduced by 1814 when the miller and corn merchant Alexander Kedslie (who left in 1928 to introduce steam power to Poland) had a dispute with the artist and near neighbour Henry Raeburn about the smoke.

The mill (as Todd's Mill) was destroyed by an explosion in 1901 when six people died. The site was then redeveloped as the Stockbridge Market and later for housing, the changes being commemorated in a plaque outside the Hamilton public house on Hamilton Terrace.

Silvermills NT248747

This hamlet lay east of Stockbridge along the line of the lade. Cumberland Hill referred to it as "ancient........with its quaint crow-stepped gables and picturesque architecture of the seventeenth century." The name appears on T. Pont's *Mercator Atlas of Lothian and Linlithgow,* (1630) and may date from the discovery of silver ore in 1607 at Hilderstan in the Bathgate Hills of West Lothian. The discovery was eagerly seized on by the hard-up monarch James VI, who promptly purchased the mines for £5,000. However the ore was smelted in Linlithgow and beside the River Avon and then shipped south from Bo'ness on the Forth to London where the king was now living, so the link with Edinburgh is dubious. Use of water power at Silvermills at some stage seems likely from the presence of the lade but there is little documentation for it beyond mention (in Shaw's *Water Power in Scotland*) of a waulker, James Forrester, operating there in 1773 and Silvermills is not mentioned in the 1850 wheel survey. A street directory for 1896 lists sawmillers, joiners and a tanner among the residents and their huddle of old houses and workshops was swept away in redevelopment in the 1990s.

Canonmills NT253749

The mills were granted to the canons of Holyrood by King David in 1128. The site was used for Scotland's second papermill in 1652-1682 by Peter Bruce, a gifted German engineer and entrepreneur, whose many activities around Scotland involved him in a series of disputes, trials and even imprisonment. Canonmills was used for grain milling by the Baxters by 1686. A stone inscribed "*the Baxters Land 1686*" was discovered during enlargement of the petrol station at the foot of Canon Street in 1973 and can be seen, together with an explanatory plaque, on the wall to the right hand side. The royal goldsmith George Heriot had the mills assigned to him, along with the rest of the Barony of Broughton, as security for a loan and much of the land is still owned by George Heriot's (School) Trust. The lade from Silvermills reached Canonmills Loch before driving (in 1850) four wheels, the largest 20 x 5 feet. The loch is on the 1817 map of Robert Kirkwood but a Gymnasium was built on the loch bed (now King George VI Park) in 1865 by John Cox of Gorgie. Photographs show a circular pond with an enormous circular rowing frame to seat over 100 rowers, plus a water ride and a tea room. The massive four-storey mill building stands at the top of Canon Street and red-roofed granaries are sited lower down.

The reconstructed undershot wheel at Bonnington. Bonnyhaugh House is behind on the left.

Logie Green (Beaverhall) Mills NT253753

These mills at the end of the lade from Dean Village are poorly recorded but appear on John Ainslie's Map of Leith (1804) and on Johnston's 1851 *Plan of Edinburgh and Leith* close to the point where the lade turned north to rejoin the river at St. Mark's Place. The name suggests an earlier waulking and bleaching site. Mills on either side of the lade in 1804 were labelled as snuff and pepper mills. In 1850 they had a single wheel (14 x 4 feet) for "ornamental stone cutting, comprising pebbles and jaspers". They lost their water power in 1890 when the Great Lade was closed but some milling of wood flour went on in the 1920s under Robert Lamb's firm of box-makers.

Bonnington Mills NT259760

An important crossing point between Newhaven and Edinburgh, Bonnington was mentioned in the charter of King David granting land and mills, the Barony of Broughton, to the

Abbot of Holyrood. In 1617 the mills were sold by the Logans of Restalrig to the City of Edinburgh and Bonnyhaugh House was built (1621) for a Dutch dyer, Jeromias Van der Heill. Grain milling continued there on the south bank of the river, accompanied by weaving, bleaching and printing. In 1882 the bleachfield area close to the weir became a skin works, latterly Messrs. Burns White, who traded until 1972. Burns Place, closeby in Newhaven Road, was erected for their workers. The weir at Redbraes supplied a long lade that drove the grain mills and then passed under the road to drive a further mill. There was a relief lade from the grain mill just below the bridge.

This downstream site (NT261761) was established in 1749-50 for a plash (yarn-washing) mill with advice from the celebrated East Lothian millwright Andrew Meikle (1719-1811) at a cost of £250. Most recently it was Inglis Paper Mill. Water power was last used in the 1930s and milling of animal food at Bonnington (T D Munro Ltd) ended in the 1960s. Edinburgh District Council eventually overcame the hard-fought objections of conservation groups and the grain mill was demolished in 1983. A restored version of the undershot wheel, dry and set in concrete, stands in Bonnyhaugh Lane as a reminder of former times (see page 71).

Leith Sawmills NT266764

Faint signs of a weir below Bowling Green Street, and the name Mill Street off Great Junction Street are the only traces of these mills. Some early maps suggest a large mill pond and the 1850 survey gives the wheels as 12 x 6 and 16 x 12 feet with only a 4 foot fall but steam working must have replaced the wheels by 1905. The mills may have been linked to the ship-building yards active on the north side of the river, with imports of timber from Scandinavia. James Scott Marshall's *Old Leith at Work* mentions records of corn mills (1722-1754) at or near to this site, suggesting a change of use from corn to timber in the mid-18th century. Earlier sawmills in Leith had been windmills.

APPENDIX 1. Sources and acknowledgements

My task has been to collate information from many sources and to relate it to my own experience of the mills and the river as they are today. This exploration of the river from one or other point of interest has occupied 30 years so far, with few weeks when my feet have not marked the riverside paths or splashed into the water.

I am extremely grateful to many people for their helpful information and advice in the writing of this little book, especially Lawrence Walker and Stanley Jamieson. The main inspiration, however, was Currie's local historian, John Tweedie, who died in 1984 without completing his definitive work on the Water of Leith and its industrial archaeology. Footnotes are unpopular with non-specialist readers so I have tried to give credit in the text where the debt is considerable and I apologise for any omissions. I thank Aberdeen University Library for permisson to reproduce George Washington Wilson's photograph of Dean Bridge and the Royal Commission on the Ancient and Historic Monuments of Scotland for the photograph of East Mill Snuff Mill, Currie. Space is short and the following list can give only the main sources of information.

> Anon (1968) **Galloways of Balerno.** Newman Neame, London.
> Coghill, Hamish (1988) **Discovering the Water of Leith.** John Donald, Edinburgh.
> Cowper, A.S (1992) **Historic Corstorphine and Roundabout.** Volume 4: From Village to Suburb. Published by the author, 32 Balgreen Road, Edinburgh.
> Cruft, Catherine (1975) **Edinburgh Old and New.** EP Publishing, East Ardsley, Yorkshire.
> Dempster, Henry (1850?) **Sketch of the Water of Leith from Buteland Farm Braes to Leith Harbour. Statistical Account of Mills and other Public Works along its banks.** Document of unknown provenance, Edinburgh Public Libraries.

Durie, Alastair J (1986) **George Washington Wilson in Edinburgh.** Kennedy Bros, Keighley, West Yorkshire.

Durie, Alastair J (1989) **Vanishing Edinburgh.** Keith Murray, Aberdeen.

Geddie, John (1896) **The Water of Leith from Source to Sea.** W H White & Co., Edinburgh.

Gladstone-Millar, Lynne (1956) **The Colinton Story.** Saint Andrew Press, Edinburgh.

Hill, Cumberland (1887) **Historic Memorials of Stockbridge, The Dean, and Water of Leith.** 2nd edition, reprinted by West Port Books, Edinburgh.

Jamieson, Stanley (editor)(1984) **The Water of Leith.** The Water of Leith Project Group, Edinburgh.

Marshall, James Scott (1977) **Old Leith at Work.** Edina Press, Edinburgh.

Shaw, Donald (1989) **The Balerno Branch and the Caley In Edinburgh.** Oakwood Press, Headington, Oxford.

Tweedie, John & Jones, Cyril (1975) **Our District.** Currie District Council.

Tweedie, John (1974) **A Water of Leith Walk, with a Historical Industrial Background, and Juniper Green its Living Centre.** Juniper Green Village Association, Edinburgh.

Vince, John (1993) **Discovering Watermills.** 6th edition. Shire Publications, Princes Risburgh.

Waterston, Robert (1945) **Early Paper Making Near Edinburgh.** Book of the Old Edinburgh Club Vol 25, 46-70.

Wenham, Peter (1989) **Watermills.** Robert Hale, London.

The **Old Ordnance Survey Maps** published by Alan Godfrey Maps, Consett, have been very useful, especially Sheets 3.03 (Inverleith & Canonmills 1896), 3.04 (Leith Walk 1894), 3.06 (Murrayfield & Blackhall 1894), 3.10 (Gorgie 1905) and 3.14 (Slateford 1893). Each map has a perceptive commentary on the district at that time by a local historian (Andrew Bethune; Barbara and R J Morris), with extracts from a contemporary Street Directory.

APPENDIX 2. Glossary of technical terms

axle: the shaft (originally wooden but later iron) on which awaterwheel (and pit wheel) rotates. Some times called the axle-tree when they were hewn from a single tree trunk.

apron weir: type of weir with strengthened riverbed downstream to counter the erosive effect of the waterfall. Redhall weir also features protruding blocks set in the face to break the force of the falling water.

Baxters: the guild of bakers. Baxter survives as a surname. Other names recalling waterside industries include Milner and Miller, Walker (Waulker), Carter and Cartwright, Wright, Kilner, Webster and Weaver, Skinner, Tanner, Dyer, Sawyer, Turner and Joiner.

bed stone: the lower, stationary, millstone.

beetle mill: finishing plant for linen cloth, pressing and polishing the cloth with rollers. c.f waulk mill for woollen cloth.

bere: a primitive variety of barley with only a few grains per stem

breast-shot: type of water wheel, hence breast wheel, where the water was supplied at about the level of the axle, as opposed to over or under the wheel.

buckets: containers for the water on an overshot or breast wheel.

bung mill: a wood-turning mill, making, among other things, bungs for beer barrels.

crown wheel: horizontal toothed wheel on the main vertical shaft of a grain mill. It transmitted the drive to ancilliary equipment, like a sack hoist (see diagram on page 23).

damhead: local name for a weir.

dressing: pattern of grooves cut into the grinding face of a mill stone.

edge-runner: type of millstone that rotated about a horizontal axis so that its grinding surface was the edge, not the face.

eye: the hole in the centre of a millstone into which the grain is fed.

fulling stocks: a set of water-driven beams arranged to fall onto wetted folded cloth to soften and thicken it.

girnal: a granary; as in the greeting "Lang may your lum reek... andmay the mouse never have cause to leave your girnal.."

grist: coarsely ground grain, either for use as cattle feed (provender) or as a first stage product for finer grinding in flour milling.

head-lade: the channel bringing water from the weir to the mill, often just called the lade, but confusingly on the Water of Leith, the mill-dam. Names for lades in other parts of Britain included lead, cut, race, flume and goit. See also tail-lade.

horse-power: The power provided by one horse, now defined as 0.75 kilowatts. The power rating of a water wheel can be calculated from its height, the flow rate from the lade (volume of water per second, which would vary with conditions) and the nominal efficiency of the type of wheel (e.g. 50% for breast wheels). The largest Water of Leith wheels were rated (nominally) at 36 h.p. or 27 kilowatts.

lint: flax, hence lintmill for a mill that processed flax for thread. Mills to crush the seed of the flax plant, linseed, (from *Linus usitatissimum*) were usually called oil mills.

overshot: type of water wheel where the water passed over the wheel, which turned clockwise (see page 9).

pitchback: type of overshot water wheel where the water was supplied to the top of the wheel but passed down the near side so that it turned anti-clockwise (see

page 9).
pit wheel: a large diameter cogwheel set vertically on the same axle as the water wheel and turning in a deep cavity or pit (see diagram on page 23).
ream: a measure (500 sheets) of paper.
retting: the process by which the soft parts of the flax plant were soaked in water and rotted away to leave the resistant fibres that could be spun into thread.
runner stone: the upper millstone, which rotated or ran.
spring water supply: By Act of Parliament in 1856 some firms were granted a supply of clean spring water direct from the Torduff treatment plant. They received (gallons/day) 81,000 (West Mill, Colinton); 72,000 (Cox's Glue Works, Gorgie); 72,000 (Kate's Mill) and 22,500 (Inglis Green Bleachfield).
spur wheel: the largest toothed cog on the main vertical shaft of a grinding mill. It conveys the drive to the cogs on secondary shafts driving the runner stones (see page 23).
tail-lade: the channel conveying water from the mill back to the river (see page 19).
thirlage: the feudal obligation of tenants to have their grain milled at the landlord's mill.
trundle: the first and lowermost horizontal cogwheel, driven by the pitwheel and attached to the main vertical drive shaft in a grinding mill. Called the wallower in England (see page 23).
undershot: type of water wheel where the water passed beneath the wheel (page 9).
waulking: process of finishing woollen cloth by rubbing and beating it in a trough of water; called fulling in England, hence fulling stocks, fuller's earth. cf beetling for linens.
weir: the wall built across the river to divert its flow into a mill lade, other common names being dam, or locally, damhead.

APPENDIX 3
Some working water mills to see in mainland Scotland.

Opening times vary with the season and checking by telephone is recommended. Some mills have a tearoom and gift shop.

Aberfeldy Water Mill. Erected in 1825 to make oatmeal and restored in 1987, this mill uses a 15 foot overshot wheel to drive two pairs of French stones, the lade running underground for 500 yards. Telephone 01887 820803.

Barry Mill, Dundee (National Trust for Scotland). Set on the small Pitairlie Burn, this 19th century oat mill is on the north side of the A930 about 7 miles west of Dundee. The mill worked commercially until the 1980s. There is a semi-circular kiln at the rear of the main mill building with the 15 foot overshot wheel on the opposite end. Grinding demonstrations are arranged from time to time as water levels and manpower allow. Telephone 01241 53321.

Blair Athol Grain Mill. Built in 1613 and restored in 1976, this was the estate mill for Athol Castle, itself a striking landmark from the nearby A9. The 16 foot breast shot wheel supplied by a lade from the River Tilt drives two pairs of French stones. Telephone 01796 481321.

Clack Mill, Kingussie. This little mill came from the Isle of Lewis in the Outer Hebrides and was reconstructed at the Highland Folk Museum in the 1940s as an example of a simple horizontal mill. The name is a variation of "Click Mill", said to suggest the sound of the mill when working. Other examples can be seen (less conveniently) at Dounby in Orkney (superb!) and Bragar on Lewis. Telephone 01540 661307.

Keathbank Mill, Blairgowrie. The River Ericht powered a series of mills here, including several on the lade downstream of the town centre. This one is upstream, alongside the Braemar Road (A93), and boasts "Scotland's largest working water wheel." plus an 1862 steam engine and other attractions. Telephone 01250 872025.

Livingston Mill Farm. Sited within a comprehensive museum at the Almond Valley Heritage Centre, West Lothian (signed off the A899), the mill dates from 1770 and has a working wheel with 3 pairs of stones for oats, corn and peas, and a kiln. A second wheel awaits restoration. Telephone 01506 414957.

Lower City Mills, Perth. This attractive pair of mills was part of a larger milling complex on the town lade in the centre of Perth. Barley and oatmeal mills shared a breast wheel and the barley mill is now a visitor centre run by Perthshire Tourist Board with the oatmeal mill working for demonstrations and a tea room in the kiln house. Telephone 01738 30572.

Montgarrie Mill, Alford, Aberdeenshire. This oatmeal mill close to the River Don is powered by a small tributary burn, still earns its living commercially and visitors can join a tour. The 25-foot overshot wheel was out of action on my visit.

New Abbey Corn Mill (Historic Scotland). A short walk along the village street from Sweetheart Abbey shows the visitor a charming renovated oatmeal mill with a working pitchback wheel, mill pond and integral kiln. About 7 miles west of Dumfries on the A710; telephone 01387 850260.

Preston Mill, East Linton, East Lothian (National Trust for Scotland). Set on the River Tyne close to the A1, Preston is arguably the prettiest of Scotland's mills, with red pantiled roofs setting off 16th and 17th century stonework (the delightful circular kiln may be even older) and a restored 13 foot breast wheel. Telephone 01620 860426.

Otterburn Mill, just south of the border in Northumberland on the A696. This is the nearest place to see a set of fulling stocks. They were originally powered by a water turbine, but the mill is now a shop for woollen and other textile products.

The new undershot wheel on the Redhall Mill lade - the first wheel to turn on the river since Bell's Mills exploded in 1972. It was built and erected in 1999 by John Parkes as a purely sentimental gesture on a workshop that occupies the site of an old piggery. The wheel is not capable of driving any machinery but is a potent reminder of what we have lost in failing to preserve any of the river's mills in working condition.

Post-script

There is still much to discover about the Water of Leith's mills and no account can ever be complete. As in scientific discovery, we are making "a series of approximations to the truth". I hope these pages will encourage someone to fill in the gaps and correct the errors in my effort, and prompt more vigorous attempts to preserve what now remains of this great industrial history.